D0498121

Phaselock Techniques

Phaselock Techniques

Floyd M. Gardner, Ph.D

Consulting Engineer
Palo Alto, California

Second Edition

A Wiley-Interscience Publication

JOHN WILEY & SONS
New York • Chichester • Brisbane • Toronto • Singapore

Copyright © 1979 by John Wiley & Sons, Inc.

All rights reserved. Published simultaneously in Canada.

Reproduction or translation of any part of this work
beyond that permitted by Sections 107 or 108 of the
1976 United States Copyright Act without the permission
of the copyright owner is unlawful. Requests for
permission or further information should be addressed to
the Permissions Department, John Wiley & Sons, Inc.

Library of Congress Cataloging in Publication Data

Gardner, Floyd Martin, 1929-
  Phaselock Techniques

  Includes bibliographies and index.
  1. Phase-locked loops. I. Title.

TK7872.P38G37   1979   621.3815′35   78-20777
ISBN 0-471-04294-3

Printed and bound in the United States of America by Braun-Brumfield, Inc.

20 19 18 17 16 15 14 13

To Sandy

Preface

This book was written for practicing engineers: for system engineers who must know the capabilities and limitations of phaselock, for equipment design engineers who build phaselock circuits, and for equipment users who must understand the operation of their equipment. For these readers the book discusses basic principles of phaselock operation, typical practices of phaselock engineering, and selected applications of phaselock to various problems.

Long mathematical derivations have been avoided on the premise that they are of little interest to the practicing engineer. Instead, I have tried to outline the underlying assumptions and methods employed in derivations and have stressed practical results. For those readers who may be interested in further details, I have listed numerous references.

On the other hand, I have avoided producing a circuits cookbook. Specific circuits quickly become obsolete; furthermore, very little fundamental understanding can be gained from a collection of circuit recipes. I have tried to stress physical understanding of basic phenomena as much as possible.

Nonetheless, because many aspects of phaselock can be expressed only in mathematical terms, the largest part of the material is presented here in that form. As a consequence, the reader must have some mathematical background. For fullest understanding of the subject some familiarity with transfer functions in the Laplace transform notation, a background in feedback or servo theory, and a nodding acquaintance with noise and spectral analysis of stochastic processes are needed. The results and applications are presented so that a less well-prepared reader can understand them, but the minimum prerequisites are necessary for a full understanding of the detailed basic principles.

This philosophy underlay the first edition of this book, which was well received by its intended audience. In the years since its publication much has happened in the phaselock world. Better analyses have been performed, new circuits have arisen (and old ones have faded away), more applications have appeared, and my own understanding of phaselock has improved substantially. Thus the time seems ripe for a new edition.

The reader will find additional analytical tools and improved explanations of the fundamentals of the subject. Moreover, the material on applications has been expanded greatly. I hope the profession finds this book to be as useful as its predecessor.

I have benefited from the wise counsel of J. L. Dautremont, Jr., W. C. Lindsey, A. J. Mallinckrodt, L. Eaton, L. M. Hershey, J. J. Spilker, Jr., J. F. Heck, C. E. Krehbiel, K. Wong, and F. Chethik, each of whom reviewed one or more chapters. Their expert critiques have made this book better than it would have been through my unaided efforts.

Floyd M. Gardner

Palo Alto, California
April 1979

Contents

Glossary of Notation

n	An integer	—
n	Loop order	14, 253
$\bar{n}, n_r$	FM click rate (clicks/sec)	187
$n(t)$	Noise voltage (V)	25, 262
$\overline{n(t)}$	Time average of noise	263
$n_c(t), n_s(t)$	Baseband quadrature components of bandpass Gaussian noise (V)	26, 268
$n'(t)$	Equivalent baseband noise added into PLL (dimensionless)	26
N, N_i	An integer representing frequency division or multiplication	—
N_o	One-sided spectral density of white noise (V^2/Hz)	29, 266
$p(n), p(\phi)$	Probability density function of $n(t)$ or ϕ	37, 263
$p(t)$	Pulse wave shape	229, 231
$P(j\omega)$	Fourier transform of pulse wave shape	229
P_s	Signal power (W)	31
$R(\tau), R_n(\tau)$	Autocorrelation function	28, 264
$s = \sigma + j\omega$	Laplace transform complex variable	9, 252
SNR	Signal-to-noise ratio	—
SNR_L	Signal-to-noise ratio in loop bandwidth $2B_L$	32
t	Time (sec)	—
T	Symbol interval of digital data stream (sec)	217, 231
T_{AV}	Average time to first cycle slip (sec)	38
T_p	Pull-in time (sec)	75
v_c, V_c	VCO control voltage (V)	8
v_d, V_d	Phase detector output voltage (V)	8, 26
V_o	Peak amplitude of VCO voltage (V)	25
V_s	Peak amplitude of signal voltage (V)	25
W_i	Spectral density of white noise (W/Hz)	31, 265
α	Limiter signal-suppression factor	126
γ	Crest factor of signal	62
θ	Phase angle (rad)	—
θ_i	Phase angle of input signal (rad)	8

θ_a	Steady-state error due to frequency ramp input;	—
	Acceleration error (rad)	45
$\theta_e = \theta_i - \theta_o$	Phase error between input signal and VCO (rad)	9, 43
θ_{no}	Fluctuation of VCO phase θ_o caused by additive noise (rad)	30
θ_o	VCO phase (rad)	8
θ_p	Loop phase error caused by oscillator noise (rad)	100
θ_v	Steady-state phase error (static phase error, loop stress) due to offset of input frequency (rad)	44
$\Delta\theta$	Phase deviation (rad)	173
$\Delta\theta$	Amplitude of phase step (rad)	44
$\Delta\theta$	Peak deviation of phase modulation (rad)	47
ζ	Damping factor of second-order loop	11
ρ	Signal-to-noise ratio	37
ρ_i	Input signal-to-noise ratio	126, 177, 226
σ_n	Standard deviation (rms value) of noise $n(t)$ (V)	263
τ	Timing error in clock synchronizer (sec)	234, 235
τ	Lag variable in computing autocorrelation (sec)	28, 264
τ	Time constant (sec)	16
τ_1, τ_2	Time constants in loop filter of second-order loop (sec)	10
ϕ	Loop phase error reduced modulo 2π (rad)	37
ϕ_n	Phase fluctuation internal to an oscillator	100
$\phi(\omega)$	Phase of transfer function of a network	254
$\Phi_n(\omega), \Phi_n(f)$	One-sided spectral density of $n(t)$. (V^2/Hz)	28, 265
ψ	Phase of a transfer function (rad)	—
$\Psi_n(\omega), \Psi(f)$	Two-sided spectral density of $n(t)$ (V^2/Hz)	265
$\omega = 2\pi f$	Angular frequency (rad/sec)	—
$j\omega$	Fourier transform variable	—
ω_i	Radian frequency of input signal (rad/sec)	25

ω_m	Modulating frequency (rad/sec)	47
ω_n	Natural frequency of second-order loop (rad/sec)	11
$\Delta\omega$	Amplitude of frequency step or of frequency offset (rad/sec)	44
$\Delta\omega$	Peak deviation of frequency modulation (rad/sec)	47
$\Delta\dot{\omega}$	Rate of change of frequency (rad/sec^2)	45
$\Delta\omega_H$	Hold-in limit of PLL (rad/sec)	53
$\Delta\omega_L$	Lock-in limit of PLL (rad/sec)	69
$\Delta\omega_P$	Pull-in limit of PLL (rad/sec)	76

Operations

$\text{Avg}(x)$	Average value—DC value—of x	—
$E(x)$	Expectation of x. Statistical average	263
$\bar{x}$	Overbar indicates averaging operation	27, 263

Phaselock Techniques

Chapter One

Introduction

1.1 NATURE OF PHASELOCK

A phaselock loop contains three basic components (Figure 1.1):

1. A phase detector (PD).
2. A loop filter.
3. A voltage-controlled oscillator (VCO), whose frequency is controlled by an external voltage.

The phase detector compares the phase of a periodic input signal against the phase of the VCO; output of the PD is a measure of the phase difference between its two inputs. The difference voltage is then filtered by the loop filter and applied to the VCO. Control voltage on the VCO changes the frequency in a direction that reduces the phase difference between the input signal and the local oscillator.

When the loop is *locked*, the control voltage is such that the frequency of the VCO is exactly equal to the average frequency of the input signal. For each cycle of input there is one, and only one, cycle of oscillator output. One obvious application of phaselock is in automatic frequency control (AFC). Perfect frequency control can be achieved by this method, whereas conventional AFC techniques necessarily entail some frequency error.

To maintain the control voltage needed for lock it is generally necessary to have a nonzero output from the phase detector. Consequently, the loop operates with some phase error present; as a practical matter, however, this error tends to be small in a well-designed loop.

A slightly different explanation may provide a better understanding of loop operation. Let us suppose that the incoming signal carries information in its phase or frequency; this signal is inevitably corrupted by additive noise. The task of a phaselock receiver is to reproduce the original signal while removing as much of the noise as possible.

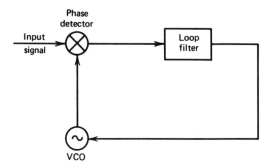

Figure 1.1   Basic phaselock loop.

To reproduce the signal the receiver makes use of a local oscillator whose frequency is very close to that expected in the signal. Local oscillator and incoming signal waveforms are compared with one another by a phase detector whose error output indicates instantaneous phase difference. To suppress noise the error is averaged over some length of time, and the average is used to establish frequency of the oscillator.

If the original signal is well behaved (stable in frequency), the local oscillator will need very little information to be able to track, and that information can be obtained by averaging for a long period of time, thereby eliminating noise that could be very large. The input to the loop is a noisy signal, whereas the output of the VCO is a cleaned-up version of the input. It is reasonable, therefore, to consider the loop as a kind of filter that passes signals and rejects noise.

Two important characteristics of the filter are that the bandwidth can be very small and that the filter automatically tracks the signal frequency. These features, automatic tracking and narrow bandwidth, account for the major uses of phaselock receivers. Narrow bandwidth is capable of rejecting large amounts of noise; it is not at all unusual for a PLL to recover a signal deeply embedded in noise.

1.2   HISTORY AND APPLICATION

An early description of phaselock was published by de Bellescize[1]* in 1932 and treated the synchronous reception of radio signals. Superheterodyne receivers had come into use during the 1920s, but there was a continual search for a simpler technique; one approach investigated was the synchronous, or homodyne, receiver.

*Superscript numbers indicate references listed at the end of each chapter.

In essence, this receiver consists of nothing but a local oscillator, a mixer, and an audio amplifier. To operate, the oscillator must be adjusted to exactly the same frequency as the carrier of the incoming signal, which is then converted to an intermediate frequency of exactly 0 Hz. Output of the mixer contains demodulated information that is carried as sidebands by the signal. Interference will not be synchronous with the local oscillator, and therefore mixer output caused by an interfering signal is a beat-note that can be suppressed by audio filtering.

Correct tuning of the local oscillator is essential to synchronous reception; any frequency error whatsoever will hopelessly garble the information. Furthermore, phase of the local oscillator must agree, within a fairly small fraction of a cycle, with the received carrier phase. In other words, the local oscillator must be phaselocked to the incoming signal.

For various reasons the simple synchronous receiver has never been used extensively. Present-day phaselock receivers almost invariably use the superheterodyne principle and tend to be highly complex. One of their most important applications is in the reception of the very weak signals from distant spacecraft.

The first widespread use of phaselock was in the synchronization of horizontal and vertical scan in television receivers[2]. The start of each line and the start of each interlaced half-frame of a television picture are signaled by a pulse transmitted with the video information. As a very crude approach to reconstructing a scan raster on the TV tube, these pulses can be stripped off and individually utilized to trigger a pair of single-sweep generators.

A slightly more sophisticated approach uses a pair of free-running relaxation oscillators to drive the sweep generators. In this way sweep is present even if synchronization is absent. Free-running frequencies of the oscillators are set slightly below the horizontal and vertical pulse rates, and the stripped pulses are used to trigger the oscillators prematurely and thus to synchronize them to the line and half-frame rates (half-frame because United States television interlaces the lines on alternate vertical scans).

In the absence of noise this scheme can provide good synchronization and is entirely adequate. Unfortunately, noise is rarely absent, and any triggering circuit is particularly susceptible to it. As an extreme, triggered scan will completely fail at a signal-to-noise ratio that still provides a recognizable, though inferior, picture.

Under less extreme conditions noise causes starting-time jitter and occasional misfiring far out of phase. Horizontal jitter reduces horizontal resolution and causes vertical lines to have a ragged appearance. Severe horizontal misfiring usually causes a narrow horizontal black streak to appear.

Vertical jitter causes an apparent vertical movement of the picture. Also, the interlaced lines of successive half-frames would so move with respect to one another that further picture degradation would result.

Noise fluctuation can be vastly reduced by phaselocking the two oscillators to the stripped sync pulses. Instead of triggering on each pulse, a phaselock technique examines the relative phase between each oscillator and many of its sync pulses and adjusts oscillator frequency so that the average phase discrepancy is small. Because it looks at many pulses, a phaselock synchronizer is not confused by occasional large noise pulses that disrupt a triggered synchronizer. The flywheel synchronizers in present-day TV receivers are really phaselocked loops. The name "flywheel" is used because the circuit is able to coast through periods of increased noise or weak signal. Substantial improvement in synchronizing performance is obtained by phaselock.

In a color television receiver, the color burst is synchronized by a phaselock loop.[3]

Spaceflight requirements inspired intensive application of phaselock methods. Space use of phaselock began with the launching of the first American artificial satellites. These vehicles carried low-power (10 mW) CW transmitters; received signals were correspondingly weak. Because of Doppler shift and drift of the transmitting oscillator, there was considerable uncertainty about the exact frequency of the received signal. At the 108-MHz frequency originally used, the Doppler shift could range over a ±3-kHz interval.

With an ordinary, fixed-tuned receiver, bandwidth would therefore have to be at least 6 kHz, if not more. However, the signal itself occupies a very narrow spectrum and can be contained in something like a 6-Hz bandwidth.

Noise power in the receiver is directly proportional to bandwidth. Therefore, if conventional techniques were used, a noise penalty of 1000 times (30 dB) would have to be accepted. (The numbers have become even more spectacular as technology has progressed; transmission frequencies have moved up to S-band, making the Doppler range some ±75 kHz, whereas receiver bandwidths as small as 3 Hz have been achieved. The penalty for conventional techniques would thus be about 47 dB.) Such penalties are intolerable and that is why narrowband, phaselocked, tracking receivers are used.

Noise can be rejected by a narrowband filter, but if the filter is fixed the signal almost never will be within the passband. For a narrow filter to be usable it must be capable of tracking the signal. A phaselocked loop is capable of providing both the narrow bandwidth and the tracking that are needed. Moreover, extremely narrow bandwidths can be conveniently obtained (3 to 1000 Hz are typical for space applications); if necessary, bandwidth is easily changed.

For a Doppler signal the information needed to determine vehicle velocity is the Doppler frequency shift. A phaselock receiver is well-adapted to Doppler recovery, for it has no frequency error when locked. (The effect of phase errors is covered in later chapters.)

1.3 OTHER APPLICATIONS

The following applications, further discussed elsewhere in the book, represent some of the current uses of phaselock.

1. One method of tracking moving vehicles involves transmitting a coherent signal to the vehicle, offsetting the signal frequency, and retransmitting back to the ground. The *coherent transponder* in the vehicle must operate so that the input and output frequencies are exactly related in the ratio m/n, where m and n are integers. Phaselock techniques are often used to establish coherence.
2. A phaselocked loop can be used as a frequency demodulator, in which service it has superior performance to a conventional discriminator.
3. Noisy oscillators can be enclosed in a loop and locked to a clean signal. If the loop has a wide bandwidth, the oscillator tracks out its own noise and its output is greatly cleaned up.
4. Frequency multipliers and dividers can be built by using PLLs.
5. Synchronization of digital transmission is typically obtained by phaselock methods.
6. Frequency synthesizers are conveniently built as phaselock loops.

1.4 PHASELOCK LITERATURE

The first edition of this book (published in 1966) contains a bibliography of about 160 entries. That was a nearly complete listing of the literature at the time it was compiled, but it quickly became outdated because of the high level of activity in the field. A 1973 bibliography[4] contained over 800 entries, so there must be over 1000 phaselock papers published by now (1978).

Clearly it is no longer practical to include a lengthy bibliography in this book, nor is it necessary. References cited in the text provide a guide to selected papers. A sampling of noteworthy papers has been reprinted in Ref. 5.

There are other books on phaselock, and the serious student ought to be aware of them. Blanchard[6] has written an introductory text that specializes in the problems of phaselocked receivers. He provides many numerical examples and graphical plots that are of value to the reader.

Viterbi has been a pioneer in the nonlinear analysis of phaselock loops; his papers have been collected in Ref. 7. Many portions of the present book owe their inspiration to his lead. The mathematical level of his approach is quite advanced, so most engineers will find learning easier from the present book or from Blanchard's book.

Massive assaults on difficult problems are found in the two works by Lindsey[8] and Lindsey and Simon.[9] A serious professional ought to be conversant with these works, but the casual user of phaselock is likely to have difficulty in extracting the information needed for ordinary engineering projects. A researcher can use these books as a gold mine of powerful analysis techniques and as a guide for extracting the last tiny measure of performance out of a PLL needed in demanding service.

The specialist should also be aware of the books by Van Trees[10] and Klapper and Frankle,[11] which relate the phaselock loop to application as a demodulator of frequency modulation.

1.5 ORGANIZATION OF BOOK

A certain amount of mathematics, network theory, and stochastic processes background is needed for an adequate understanding of phaselock and therefore a very brief summary and review of the pertinent material is given in Appendix A.

Fundamentals of phaselock are presented in Chapters 2 to 7 and applications are found in Chapters 8 to 11. The basic notion of transfer function of a linear loop is introduced in Chapter 2, along with definitions of loop parameters and identification of important configurations of practical loops.

Noise, in both linear and nonlinear regimes, is expounded in Chapter 3. Tracking performance is discussed in Chapter 4 and methods of bringing the loop into lock are found in Chapter 5.

Implementation problems in the loop components, namely loop filter, phase detector, and VCO, are discussed in Chapter 6. The design engineer ought to become familiar with this material before attempting any but the simplest of circuits.

Loop optimization is treated in Chapter 7, as well as in other portions of the book.

Chapter 8 is devoted to some of the problems that arise in phaselocked receivers and other long-loop configurations. Phaselocked modulators and demodulators are treated in Chapter 9, locked oscillators and synthesizers are the subject of Chapter 10, and a brief survey of the phaselock aspects of synchronizers for digital data is given in Chapter 11.

REFERENCES

1. H. de Bellescize, "La Reception Synchrone," *Onde Electr.*, Vol. 11, pp. 230–240, June 1932.
2. K. R. Wendt and G. L. Fredendall, "Automatic Frequency and Phase Control of Synchronization in Television Receivers," *Proc. IRE*, Vol. 31, pp. 7–15, January 1943.
3. D. Richman, "Color-Carrier Reference Phase Synchronization in NTSC Color Television," *Proc. IRE*, Vol. 42, pp. 106–133, January 1954.
4. W. C. Lindsey and R. C. Tausworthe, *A Bibliography of the Theory and Application of the Phase-Lock Principle*, Technical Report 32–1581, Jet Propulsion Laboratory, Pasadena, CA, April 1, 1973.
5. W. C. Lindsey and M. K. Simon (Eds.), *Phaselocked Loops and Their Application*, IEEE Press, New York, 1978.
6. A. Blanchard, *Phase-Locked Loops: Application to Coherent Receiver Design*, Wiley, New York, 1976.
7. A. J. Viterbi, *Principles of Coherent Communication*, McGraw-Hill, New York, 1966.
8. W. C. Lindsey, *Synchronization Systems in Communication and Control*, Prentice-Hall, Englewood Cliffs, NJ, 1972.
9. W. C. Lindsey and M. K. Simon, *Telecommunication Systems Engineering*, Prentice-Hall, Englewood Cliffs, NJ, 1973.
10. H. L. Van Trees, *Detection, Estimation and Modulation Theory*: Part II, Wiley, New York, 1971.
11. J. Klapper and J. T. Frankle, *Phaselocked and Frequency Feedback Systems*, Academic Press, New York, 1972.

Chapter Two

Loop Fundamentals

2.1 BASIC TRANSFER FUNCTIONS

Let us consider an elementary loop consisting of a phase detector (PD), a loop filter, and a voltage-controlled oscillator (VCO) as in Figure 2.1. The graphical symbols of the figure are frequently encountered and are used throughout this book.

The input signal has a phase of $\theta_i(t)$, and the VCO output has a phase $\theta_o(t)$. For the present we assume that the loop is locked, that the phase detector is linear and that the PD output voltage is proportional to the difference in phase between its inputs; that is,

$$v_d = K_d(\theta_i - \theta_o) \qquad (2.1)$$

where K_d is called the *phase-detector gain factor* and is measured in units of volts per radian.

Phase error voltage v_d is filtered by the loop filter. Noise and high-frequency signal components are suppressed; also, the filter helps to determine dynamic performance of the loop. Filter transfer function is given by $F(s)$.

Frequency of the VCO is determined by the control voltage v_c. Deviation of the VCO from its center frequency is $\Delta\omega = K_o v_c$ where K_o is the *VCO gain factor* and has units of rad/sec-V. Since frequency is the derivative of phase, the VCO operation may be described as $d\theta_o/dt = K_o v_c$. By taking Laplace transforms we obtain

$$L\left[\frac{d\theta_o(t)}{dt}\right] = s\theta_o(s) = K_o V_c(s) \qquad (2.2)$$

therefore,

$$\theta_o(s) = \frac{K_o V_c(s)}{s}$$

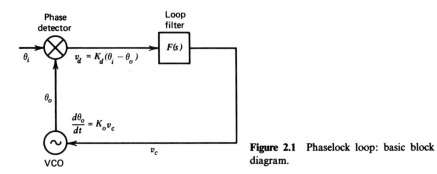

Figure 2.1 Phaselock loop: basic block diagram.

In other words, the phase of the VCO output is linearly related to the integral of the control voltage.

By using Laplace notation* the following equations are applicable:

$$V_d(s) = K_d[\theta_i(s) - \theta_o(s)] \qquad (2.3)$$

$$V_c(s) = F(s)V_d(s) \qquad (2.4)$$

$$\theta_o(s) = \frac{K_o V_c(s)}{s} \qquad (2.5)$$

Combination of these equations results in the basic loop equations

$$\frac{\theta_o(s)}{\theta_i(s)} = H(s) = \frac{K_o K_d F(s)}{s + K_o K_d F(s)} \qquad (2.6)$$

$$\frac{\theta_i(s) - \theta_o(s)}{\theta_i(s)} = \frac{\theta_e(s)}{\theta_i(s)} = \frac{s}{s + K_o K_d F(s)} = 1 - H(s) \qquad (2.7)$$

$$V_c(s) = \frac{s K_d F(s) \theta_i(s)}{s + K_o K_d F(s)} = \frac{s \theta_i(s)}{K_o} H(s) \qquad (2.8)$$

where $H(s)$ is the *closed-loop transfer function*.

Before proceeding further, it is necessary to specify the loop filter $F(s)$.

2.2 SECOND-ORDER LOOP

Two widely used loop filters are shown with their respective transfer functions in Figure 2.2. The passive filter is quite simple and is often

*Notation example: $V_d(s) \triangleq L[v_d(t)]$. Similar relations apply to the other quantities.

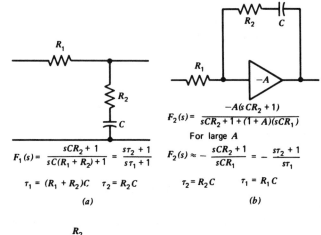

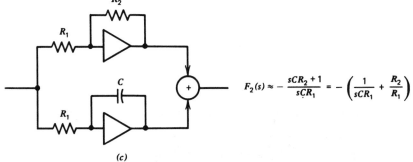

Figure 2.2 Filters used in second-order loop: (*a*) passive filter; (*b*) active filter; (*c*) alternative active filter. Both active filters have the same transfer function. The first active circuit is the one most often used, but the alternative form is sometimes more convenient.

satisfactory for many purposes. The active filter requires a high-gain DC amplifier but provides better tracking performance, as is shown in Chapter 4.

For the passive filter the closed-loop transfer function is

$$H_1(s) = \frac{K_o K_d(s\tau_2 + 1)/\tau_1}{s^2 + s(1 + K_o K_d \tau_2)/\tau_1 + K_o K_d/\tau_1} \tag{2.9}$$

For the active filter, after accommodating the phase reversal of the amplifier, the closed-loop transfer function is found to be

$$H_2(s) = \frac{K_o K_d(s\tau_2 + 1)/\tau_1}{s^2 + s(K_o K_d \tau_2/\tau_1) + K_o K_d/\tau_1} \tag{2.10}$$

provided that amplifier gain is very large. These transfer functions may be rewritten as

$$H_1(s) = \frac{s\left(2\zeta\omega_n - \omega_n^2/K_oK_d\right) + \omega_n^2}{s^2 + 2\zeta\omega_n s + \omega_n^2} \qquad (2.9a)$$

$$H_2(s) = \frac{2\zeta\omega_n s + \omega_n^2}{s^2 + 2\zeta\omega_n s + \omega_n^2} \qquad (2.10a)$$

in which, drawing on servo terminology, ω_n is the *natural frequency* of the loop and ζ is the *damping factor*.

Passive filter	Active filter	
$\omega_n = \left(\dfrac{K_oK_d}{\tau_1}\right)^{1/2}$	$\omega_n = \left(\dfrac{K_oK_d}{\tau_1}\right)^{1/2}$	
$\zeta = \dfrac{1}{2}\left(\dfrac{K_oK_d}{\tau_1}\right)^{1/2}\left(\tau_2 + \dfrac{1}{K_oK_d}\right)$	$\zeta = \dfrac{\tau_2}{2}\left(\dfrac{K_oK_d}{\tau_1}\right)^{1/2} = \dfrac{\tau_2\omega_n}{2}$	(2.11)
$\tau_1 = (R_1 + R_2)C$	$\tau_1 = R_1C$	
$\tau_2 = R_2C$	$\tau_2 = R_2C$	

We observe that the two transfer functions are nearly the same if $1/K_oK_d \ll \tau_2$ in the passive filter. It becomes apparent later that the active filter is close to an ideal towards which we strive and that the passive filter is an imperfect imitation.

Because the highest power of s in the denominator of the transfer function is 2, the loop is known as *second-order loop*. This form of second-order loop is widely applied because of its simplicity and good performance.

According to servo terminology, the *type* of a loop is a number equal to the number of perfect integrators within the loop. Any PLL is at least a Type I loop because of the perfect integrator in the VCO. If the loop filter contains one perfect integrator, then the loop is Type II; a second-order PLL with a high-gain active filter approximates a Type II loop, whereas a PLL with passive filter is Type I. The alternative form of active filter in Figure 2.2 emphasizes the presence of the integrator.

The magnitude of the frequency response of a high-gain loop for several values of damping factor is plotted in Figure 2.3. It can be seen that the loop performs a lowpass filtering operation on phase inputs.

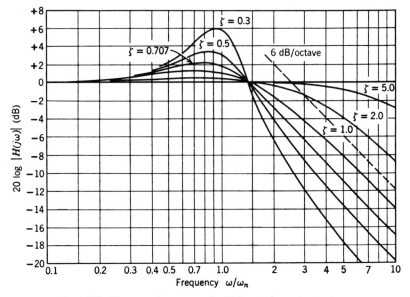

Figure 2.3 Frequency response of a high-gain second-order loop.

Error response of the loop is also of interest. For a high-gain,* second-order loop the error response is

$$\frac{\theta_e(s)}{\theta_i(s)} = \frac{s^2}{s^2 + 2\zeta\omega_n s + \omega_n^2} \tag{2.12}$$

whereas for a low-gain loop

$$\frac{\theta_e(s)}{\theta_i(s)} = \frac{s(s + \omega_n^2/K_o K_d)}{s^2 + 2\zeta\omega_n s + \omega_n^2} = \frac{s[s + 1/\tau_1]}{s^2 + 2\zeta\omega_n s + \omega_n^2} \tag{2.13}$$

Error response is plotted in Figure 2.4 for a high-gain loop with $\zeta = 0.707$. A highpass characteristic is obtained; that is, the loop tracks low-frequency changes but cannot track high frequencies.

The transfer function $H(s)$ has a well-defined 3-dB bandwidth, which we label ω_{3dB}. There is generally very little reason to be interested in ω_{3dB} of a PLL, but its relation to ω_n is presented here to provide a comparison with a familiar concept of bandwidth. By setting $|H(j\omega)|^2 = 0.5$ and solving for ω

---

*A high-gain loop has either a passive filter with $K_o K_d \tau_2 \gg 1$ or an active filter with $|A|\tau_1 \gg \tau_2$.

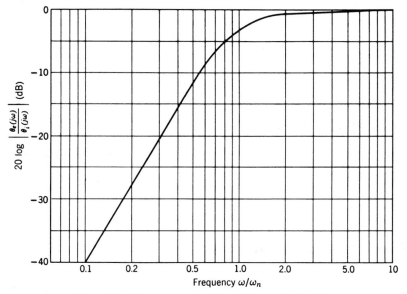

Figure 2.4   Error response of high-gain loop, $\zeta = 0.707$.

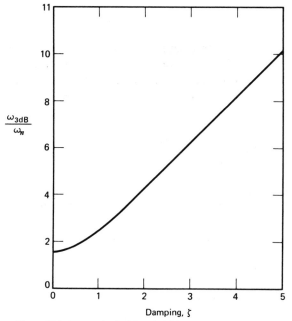

Figure 2.5   Three-decibel bandwidth of a second-order loop.

we find that

$$\omega_{3dB} = \omega_n \left[2\zeta^2 + 1 + \sqrt{(2\zeta^2 + 1)^2 + 1} \ \right]^{1/2}$$

Typical values are plotted in Figure 2.5 for a high-gain loop.

2.3 LOOP GAIN; NOTATION

The open-loop transfer function (open-loop gain) of any PLL, not just one of second order, is

$$G(s) = \frac{K_o K_d F(s)}{s} \qquad (2.14)$$

whereby the closed-loop transfer function is

$$H(s) = \frac{G(s)}{1 + G(s)} \qquad (2.15)$$

We define the *DC gain* of the loop as

$$K_v = K_o K_d F(0) \qquad (2.16)$$

which has dimensions of frequency (i.e., time^{-1}) in radians per second. (A reason for choosing the subscript v is given in Chapter 4.) It is demonstrated later that good performance of the loop usually requires a large value for K_v. Transfer function $F(s)$ of the loop filter is a rational function of the form

$$F(s) = \frac{g(s - z_1)(s - z_2) \cdots (s - z_m)}{(s - p_1)(s - p_2)(s - p_3) \cdots (s - p_{n-1})}$$

for an nth-order PLL. If the filter is to be realizable, m cannot exceed $n - 1$ (Appendix A). The factor g is a multiplicative constant. If $m = n - 1$ (as occurs often in PLLs), then $g = F(\infty)$.

When $F(s)$ is expanded in partial fractions, the open-loop gain may be written as

$$G(s) = \frac{K}{s} \left[a_1 + \sum_{i=1}^{n-1} \frac{a_{i+1}}{s - p_i} \right] \qquad (2.17)$$

assuming, temporarily, that all poles are simple with residues a_{i+1}. We choose to give K the name of *loop gain*; K has the dimensions of radian frequency and is shown presently to be an important parameter of a PLL. The dimensionless quantity a_1 is zero if m is less than $n-1$, while $a_1 \triangleq 1$ if $m = n - 1$.

The above definition of K is unambiguous only if $a_1 = 1$. If $a_1 = 0$ the rules are incomplete and K is not properly defined. The most important PLLs are designed for $a_1 = 1$ (that is, equal number of poles and zeros in the loop filter). Rather than attempt a universal definition of K, it proves more convenient to treat other conditions case by case. Judicious definition of K provides a simple basis for comparing loops with different forms of transfer functions.

Examples

1. Second-order PLL, passive lag–lead filter

$$F(s) = \frac{s\tau_2 + 1}{s\tau_1 + 1}$$

$$G(s) = \frac{K_o K_d \tau_2}{\tau_1 s}\left(1 + \frac{1/\tau_2 - 1/\tau_1}{s + 1/\tau_1}\right) = \frac{K}{s}\left(1 + \frac{a_2}{s + 1/\tau_1}\right)$$

$$K = \frac{K_o K_d \tau_2}{\tau_1}; \quad a_1 = 1; \quad a_2 = \frac{1}{\tau_2} - \frac{1}{\tau_1}$$

2. Second-order PLL, active filter

$$F(s) = \frac{s\tau_2 + 1}{s\tau_1}$$

$$G(s) = \frac{K_o K_d \tau_2}{s\tau_1}\left(1 + \frac{1}{s\tau_2}\right) = \frac{K}{s}\left(1 + \frac{a_2}{s}\right)$$

$$K = \frac{K_o K_d \tau_2}{\tau_1}; \quad a_1 = 1; \quad a_2 = \frac{1}{\tau_2}$$

$$\omega_n = \sqrt{K a_2}; \quad \zeta = \frac{1}{2}\sqrt{\frac{K}{a_2}}$$

3. An nth-order PLL with all open-loop poles at $s=0$ and $n-1$ arbitrary zeros

$$G(s) = \frac{K}{s}\left(1 + \frac{a_2}{s} + \frac{a_3}{s^2} + \cdots + \frac{a_n}{s^{n-1}}\right)$$

$$H(s) = \frac{K\left(s^{n-1} + a_2 s^{n-2} + \cdots a_n\right)}{s^n + K\left(s^{n-1} + a_2 s^{n-2} + \cdots a_n\right)}$$

The last example illustrates how to deal with multiple, coincident poles at the origin.

2.4 OTHER LOOP ORDERS

The reader should not conclude that second-order is the only loop order that may be used. It is certainly the most prevalent, but there are applications in which some other loop order is acceptable or even necessary.

A first-order loop is obtained if the filter is omitted, that is, $F(s) = 1$. Loop gain is simply $K = K_o K_d = K_v$ and $a_1 = 1$. The closed-loop transfer function is

$$H(s) = \frac{K}{s + K} \tag{2.18}$$

so the loop gain is the only parameter available to the designer. The first-order loop has a 3-dB bandwidth of K rad/sec. If it is necessary to have large DC gain (often needed to ensure good tracking), then the bandwidth must also be large. Therefore, narrow bandwidth and good tracking are incompatible in the first-order loop; for this reason it is not often used.

Another simple filter is provided by an RC lag network (Figure A.4), which has the transfer function

$$F(s) = \frac{1}{s\tau + 1} \tag{2.19}$$

From the previous section we note that $a_1 = 0$, so the rules do not directly provide a definition of K. It is convenient to define $K = K_o K_d$, just as in a first-order loop. Then $a_2 = 1/\tau$ and the loop transfer function is

$$H(s) = \frac{Ka_2}{s^2 + a_2 s + Ka_2} \tag{2.20}$$

The denominator is of second degree, so, strictly speaking, the loop qualifies as a second-order loop. In later chapters it becomes apparent that this particular type of loop might best be regarded as a modified first-order loop, despite the presence of two poles in its transfer function.

We can define $\omega_n = \sqrt{Ka_2}$ and $\zeta = \frac{1}{2}\sqrt{a_2/K}$ and obtain the loop transfer function

$$H(s) = \frac{\omega_n^2}{s^2 + 2\zeta\omega_n s + \omega_n^2} \qquad (2.21)$$

There are two circuit parameters available (τ and K), whereas usually three loop parameter specifications must be met (ω_n, ζ, and K_v). Obviously, the three loop parameters cannot be chosen independently. If it is necessary to have large DC gain and small bandwidth, the loop will be badly underdamped and transient response will be poor.

A very similar condition is found in servomechanisms; in the simplest servos damping becomes very small as gain increases. The solution to the servo problem is to employ tachometer feedback or to use *lag–lead* compensation. The latter expedient is commonly used in phaselock loops and results in the filters of Figure 2.2, which were analyzed earlier. Because the lag–lead filter has two independent time constants, the natural frequency and damping can be chosen independently. Furthermore, DC gain can be made as large as may be necessary for good tracking.

The lag pole-frequency of an active filter approaches zero and practically is somewhat vague, since amplifier gain is likely to have large tolerances (see Section 6.1). A better description for the active filters of Figure 2.2 might be *proportional-plus-integral control*, another term borrowed from the servo field. The configurations of Figures 2.2*a* and 2.2*b* emphasize the lag–lead character of the loop filter of a second-order loop, while Figure 2.2*c* emphasizes the proportional-plus-integral control character. Both viewpoints are equally valid.

In a passive filter, we necessarily have $\tau_1 > \tau_2$, but no such restriction exists in an active filter. It is possible and reasonable to have gain, rather than attenuation, in the active loop filter at high frequencies.

There are situations in which a third-order loop provides useful performance characteristics not obtainable with a simpler loop. Accordingly, it is sometimes used in special applications.

It is seen in Chapter 4 that the third-order loop is most useful if the two poles of its filter are at the origin, that is, if the loop filter contains two cascaded integrators. In this case the transfer function of the loop is (using the notation of Section 2.3)

$$H(s) = \frac{K(s^2 + a_2 s + a_3)}{s^3 + K(s^2 + a_2 s + a_3)} \qquad (2.22)$$

Complex zeros are permitted as is illustrated in Chapter 7.

It is rare that a loop is constructed with an order higher than third. One reason is that there has been little need for higher-order loops in the applications in which phaselock techniques are most commonly employed. Also, the closed-loop parameters of high-order, active networks tend to be overly sensitive to changes of gain and circuit components. Finally, it is more difficult to stabilize a high-order loop, whereas the second-order loop, as commonly built, is unconditionally stable. (Parasitic circuit elements often cause a loop intended to be second order to be actually of much higher order. It is usually expedient to treat such a loop as basically second order and handle the parasitic effects as perturbations. Examples are given in Chapter 8.)

2.5   ROOT-LOCUS PLOTS

Considerable insight into the behavior of a phaselocked loop can be attained by determining the locations of the poles of the closed-loop response. These poles change their locations as the loop gain is changed. The path that the poles trace out in their migrations in the complex s-plane is known as the *root-locus plot*. A major advantage of the root-locus method is that the plot can be determined graphically and relatively quickly by working solely from the locations of the known open-loop poles and zeros and utilizing a few simple rules.*

Typically, the locus is drawn for the full range of gain variation, from zero to infinity. The plot starts (zero gain) on the open-loop poles and terminates (infinite gain) on the open-loop zeros (some of which may be located at infinity). Open-loop transfer function for any PLL is

$$G(s) = \frac{K_o K_d F(s)}{s} \qquad (2.23)$$

Thus open-loop poles always include one at the origin besides the poles of $F(s)$. The open-loop zeros are the zeros of $F(s)$ and a zero at infinity due to the $1/s$ term.

---

*For an extensive description of root-locus methods see J. G. Truxal, *Automatic Feedback Control System Synthesis*, McGraw-Hill, New York, 1955. Chap. 4.

As might be expected, the first-order loop $[F(s)\equiv 1]$ has the simplest root locus. There is a single open-loop pole at the origin and a single zero at infinity, and the closed-loop pole moves along the negative real axis from zero to infinity as the gain increases.

A loop that uses only a lag filter $F(s)=(s\tau+1)^{-1}$ has two open-loop poles, one at zero and one at $s=-1/\tau$, and two zeros at infinity. The root locus is sketched in Figure 2.6a. Initially, the poles move toward each other on the negative real axis. When they meet halfway, they become a complex conjugate pair and move toward infinity along a vertical line at $s=-1/2\tau$. It may be seen that damping becomes very small as gain increases.

The benefit obtainable from a lead term may be seen in Figure 2.6b. Here, too, the poles migrate together and become complex when they

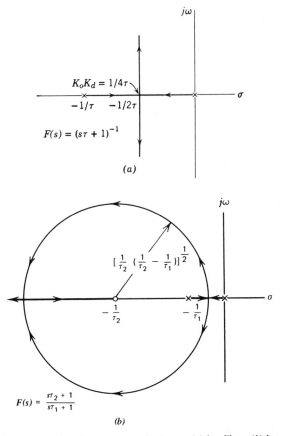

Figure 2.6 Root-locus plots for second-order loops: (a) lag filter; (b) lag–lead filter.

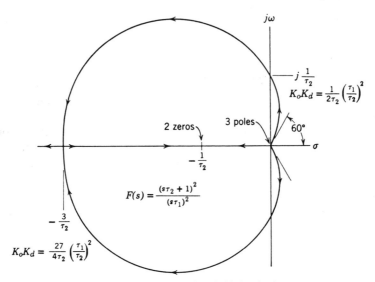

Figure 2.7   Root locus plot of third-order loop.

meet. However, because of the finite zero, the complex portion of the locus is now a circle centered at $-1/\tau_2$. Damping is small for moderately small gains, but beyond a minimum point damping increases with increasing gain. With sufficiently high gain, the locus eventually returns to the real axis and the loop is overdamped. One branch of the locus terminates at the finite zero; the other terminates at infinity.

The plot of Figure 2.6*b* is for a passive loop filter. If the loop filter were a perfect integrator (an idealized active filter), both poles would originate at the origin, and the radius of the circle would be $1/\tau_2$. Otherwise the plot would be little altered.

It is common practice to design second-order loops with a damping factor $\zeta = 1/\sqrt{2} = 0.707$, in which case the closed-loop poles lie on radial lines located at $\pm 45°$ from the negative real axis. With an active filter, for which the root-locus circle passes through the origin of the *s*-plane, these poles are located at $(-1 \pm j1)/\tau_2$. Thus the poles are located on the same vertical line as the zero; furthermore, $\omega_n = \sqrt{2}/\tau_2$. Similarly, if $\zeta = 0.5$, it turns out that $\omega_n = 1/\tau_2$, and a circle of radius $1/\tau_2$ centered at the origin passes through both poles and the zero. If $\zeta = 1$, the poles are coincident on the negative real axis and $\omega_n = 2/\tau_2$.

A third-order loop has two zeros and two poles in its loop filter, which leaves four parameter choices open to the designer. It is shown in Chapter 4 that the third-order loop is most useful if both filter poles are at the origin. Purely for convenience we also assume that the zeros are coincident at $s = -1/\tau_2$. Figure 2.7 shows the root locus of this specific loop. Because it is especially easy to compute and plot, this particular locus was chosen for illustration[*]; however, the general characteristics of the plot are fairly typical of any third-order loop that would be considered useful.

One feature of the plot is striking; the locus enters the right half-plane for low values of gain, and the loop is unstable for that condition. This is in direct contrast to the first- and second-order loops, which were unconditionally stable for all values of gain. When a third-order loop is used, the gain must be prevented from falling into the unstable region.

2.6 BODE PLOTS

Another useful tool in the study of PLLs is the *Bode plot*, which is a pair of graphs displaying the polar components of $G(j\omega)$, the open-loop transfer function (see Appendix A). Several loop parameters appear as distinctive points on the graphs. Also, Bode plots are well suited for experimental analysis of loop stability (see Chapter 8).

The Bode plot for a first-order loop is shown in Figure 2.8. The only frequency-selective term arises from the integration action of the VCO; the magnitude plot is a straight line (on the log–log scale) with slope of -6 dB/octave and the phase is constant at $-90°$. Since a VCO is present in every PLL, the Bode plot of the VCO is embedded in the plot of any higher-order loop.

Gain crossover (the frequency at which $|G| = 1$, i.e., 0 dB) of the first-order loop occurs at $\omega = K$, the loop gain of Section 2.3. The straight line and its crossover completely define the linear dynamics of the first-order loop.

Insertion of a simple lag filter (2.19) into the loop causes a break in the magnitude curve to -12 dB/octave for frequencies above $\omega = 1/\tau$, as in Figure 2.9. The break usually is placed at a frequency beyond crossover so as to obtain a satisfactory value of damping. If the break is at crossover, then damping is $\zeta = 0.5$. If the break is at a frequency below crossover, damping is less than 0.5: a condition that one ordinarily tries to avoid.

Gain crossover of the -6 dB/octave line segment (or its extension) occurs at $\omega = K$. Phase is $-90°$ at low frequencies but approaches $-180°$ at high frequencies. The additional phase lag is $45°$ at the break frequency.

---

[*]Furthermore, several designers of third-order loops have chosen this configuration in their designs.

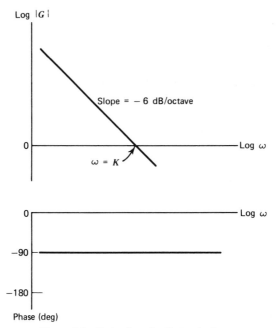

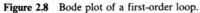

Figure 2.8 Bode plot of a first-order loop.

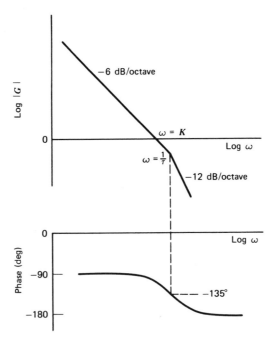

Figure 2.9 Bode plot of a loop with lag filter; $F(s)=(s\tau+1)^{-1}$.

Bode plots for the important second-order loops are shown in Figure 2.10. At very low frequencies, the VCO integration is dominant so the amplitude slope is -6 dB/octave and the phase is $-90°$. The pole of the loop filter introduces another lag: at $\omega \cong 1/A\tau_1$ for an active filter and at $\omega = 1/\tau_1$ for a passive filter. Slope becomes -12 dB/octave and phase approaches $-180°$

If the gain of the active filter were infinite—if the filter contained an ideal integrator—the filter pole would be at zero frequency, so the slope would start at -12 dB/octave and the phase at $-180°$. It is often convenient and justified to aproximate the gain as infinite and to draw the Bode plot accordingly.

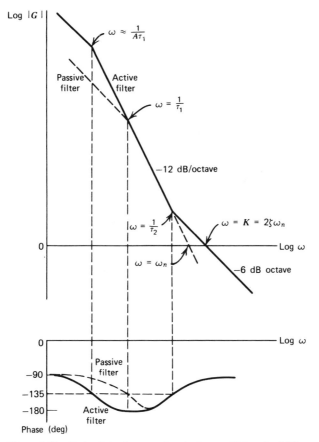

Figure 2.10 Bode plot of a second-order loop with lag–lead filter.

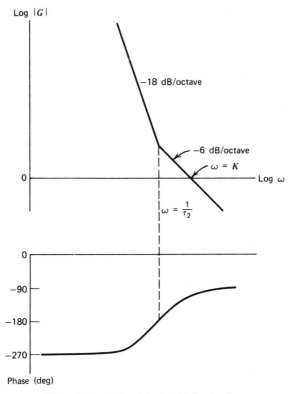

Figure 2.11 Bode plot of a third-order loop.

The stabilizing zero introduces a lead that causes the slope to revert to -6 dB/octave and the phase to approach $-90°$ for high frequencies. The break in slope occurs at $\omega = 1/\tau_2$.

Natural frequency ω_n is the frequency at which the extension of the -12 dB/octave segment crosses the unity-gain ordinate. Placing the lead break at the unity-gain point yields damping $\zeta = 0.5$. Since smaller damping is rarely wanted, the break almost invariably is placed above unity gain

Crossover of the final -6 dB/octave segment (or its extension if damping should be less than 0.5) occurs at a frequency $\omega = K = 2\zeta\omega_n$. The parameter K appears often in the chapters that follow.

As a last example, Figure 2.11 shows the Bode plot for the third-order loop whose root locus is shown in Figure 2.7. Because the loop filter now contains two ideal integrators, the low-frequency slope is -18 dB/ octave and the phase is $-270°$. Two lead zeros are needed to break the slope to -6 dB/octave at high frequencies; the zeros are arbitrarily shown as coincident. Gain crossover again occurs at $\omega = K$.

Chapter Three

Noise Performance

A major attraction of the phaselock loop is its ability to cope with large amounts of noise. Analyses of the effects of noise are presented in this chapter.

3.1 LINEAR ANALYSIS

We consider the phase detector to be a perfect multiplier with two inputs $v_1(t)$ and $v_2(t)$. Its output is $K_m v_1 v_2$, where K_m is a constant with dimensions of volts^{-1}. (Phase detectors are frequently modeled as multipliers, partly for analytical convenience and partly because many practical phase detectors are good approximations to multipliers; see Chapter 6.)

We let one input to the multiplier consist of a sinusoidal signal plus stationary, gaussian, bandpass noise:

$$v_i(t) = V_s \sin(\omega_i t + \theta_i) + n(t) \tag{3.1}$$

The other input to the multiplier comes from the VCO* and has the form

$$v_o(t) = V_o \cos(\omega_i t + \theta_o) \tag{3.2}$$

For purposes of this chapter the input phase θ_i is assumed to be time invariant. The effects of time-dependent θ_i are presented in the next chapter.

Treatment of θ_o is somewhat less straightforward. Temporarily we also assume θ_o to be time invariant, but that condition clearly does not occur in a real loop. Noise accompanying the signal causes the VCO phase to

---

*Note that v_i and v_o are really 90° out of phase with one another. The input has been written as a sine and the VCO voltage has been written as a cosine. The two phases θ_i and θ_o are based on these quadrature references. It is typical of multiplier-type phase detectors that the VCO locks in quadrature to the incoming signal, so the notation is arranged in anticipation of this fact.

fluctuate; determining the statistics of those fluctuations is the objective of this section.

To proceed we assume a fictitious open-loop condition whereby noise cannot reach the VCO. Furthermore, VCO frequency is assumed to be exactly equal to the signal frequency and θ_o is assumed to be time invariant, though arbitrary. In essence, we restrict attention to the phase detector alone for the first part of the analysis. Later, the loop is closed and time-dependent θ_o is admitted.

The bandpass input noise can be expanded (Appendix A) into two quadrature, independent components to give

$$n(t) = n_c(t)\cos\omega_i t - n_s(t)\sin\omega_i t \tag{3.3}$$

whereupon the output of the multiplier is found to be

$$v_d(t) = K_m v_i(t) v_o(t)$$

$$= \tfrac{1}{2} K_m V_s V_o \sin(\theta_i - \theta_o) + \tfrac{1}{2} K_m n_c V_o \cos\theta_o + \tfrac{1}{2} K_m V_o n_s \sin\theta_o$$

$$+ \tfrac{1}{2} K_m V_s V_o \sin(2\omega_i t + \theta_i + \theta_o) + \tfrac{1}{2} K_m V_o n_c \cos(2\omega_i t + \theta_o)$$

$$- \tfrac{1}{2} K_m V_o n_s \sin(2\omega_i t + \theta_o) \tag{3.4}$$

The product consists of three low-frequency terms and three terms at twice the input frequency. Our interest is in the difference-frequency terms, so the double-frequency terms are discarded for this analysis.

In practice, filtering or other expedients must be applied to suppress the double-frequency *ripple*. It is ignored here, but ripple is a serious disturbance in many applications and substantial effort is often needed to eliminate it. Chapter 6 contains some examples of ripple-reduction techniques.

Now we define $K_d = \tfrac{1}{2} K_m V_s V_o$ (justification is provided presently), so the multiplier output, after the ripple is discarded, becomes

$$v_d = K_d \sin(\theta_i - \theta_o) + \frac{n_c K_d}{V_s}\cos\theta_o + \frac{n_s K_d}{V_s}\sin\theta_o \tag{3.5}$$

Next we define $n'(t)$ as

$$n'(t) = \frac{n_c(t)}{V_s}\cos\theta_o + \frac{n_s(t)}{V_s}\sin\theta_o \tag{3.6}$$

which is a dimensionless quantity [as opposed to $n(t)$, which has dimensions of volts].

Output of the phase detector is thereby simplified to

$$v_d = K_d[\sin(\theta_i - \theta_o) + n'(t)] \qquad (3.7)$$

An exact equivalent circuit of the phase detector is shown in Figure 3.1. The phase detector output consists of the linear superposition of a signal term $K_d \sin(\theta_i - \theta_o)$ and a noise term $K_d n'(t)$. No linearizing approximation has yet been imposed.

The signal-linearizing approximation invariably made requires that $(\theta_i - \theta_o)$ be small so that $\sin(\theta_i - \theta_o) \simeq \theta_i - \theta_o$ and the useful output of the phase detector is approximated by $K_d(\theta_i - \theta_o)$, as in Chapter 2. We note that K_d is proportional to input-signal level. Therefore, if the input-signal amplitude varies, K_d and all loop parameters dependent on loop gain also vary.

Linearization is necessary to permit analysis in terms of transfer functions and spectral densities. Note that removal of the sinusoidal signal nonlinearity has no influence on the equivalent noise $n'(t)$. Linearization with respect to noise proves to be more subtle.

But first, let us develop some of the statistical properties of $n'(t)$. From Appendix A, it can be concluded that n' has zero mean. (Explicit dependence on t is dropped for notational convenience.) If θ_o is assumed to be time invariant, though arbitrary, the variance of n' is

$$\sigma_{n'}^2 = \frac{1}{V_s^2}\left(\overline{n_c^2}\cos^2\theta_o + \overline{n_s^2}\sin^2\theta_o + 2\overline{n_c n_s}\sin\theta_o\cos\theta_o\right) \qquad (3.8)$$

where the overbar symbolizes the averaging operation.

From Appendix A we have $\overline{n_c^2} = \overline{n_s^2} = \overline{n^2} = \sigma_n^2$ and $\overline{n_c n_s} = 0$. Moreover, $\cos^2\theta_o + \sin^2\theta_o = 1$, so

$$\sigma_{n'}^2 = \frac{\sigma_n^2}{V_s^2} \qquad (3.9)$$

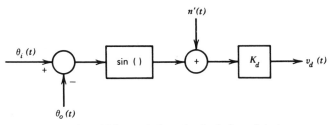

Figure 3.1 Noise equivalent circuit of phase detector.

Intensity of the equivalent noise is rotationally invariant; it does not depend on the value of θ_o. [The same result obtains if θ_o is time varying, provided $\theta_o(t)$ is independent of $n'(t)$. Independence is assured if the feedback loop is open; however in the normal, closed-loop condition, noise clearly affects VCO phase, so the two cannot be strictly independent. It has been usual to assume that θ_o and n' are approximately independent, provided that the bandwidth* of the phaselock loop is much smaller than the bandwidth of the input noise. This is a restrictive assumption that is avoided here. In consequence, the result of our analysis will apply to any arbitrary noise-spectrum and is not confined to white-noise conditions.]

Now we examine the spectrum of n', which is obtained by finding the autocorrelation function and taking the Fourier transform (Appendix A). From (3.6), the autocorrelation of n' is

$$\overline{n'(t_1)n'(t_2)} = \frac{1}{V_s^2}\left\{\cos^2\theta_o\,\overline{n_c(t_1)n_c(t_2)} + \sin^2\theta_o\,\overline{n_s(t_1)n_s(t_2)}\right.$$

$$\left. + \sin\theta_o\cos\theta_o\left[\,\overline{n_c(t_1)n_s(t_2)} + \overline{n_s(t_1)n_c(t_2)}\,\right]\right\} \quad (3.10)$$

But, from Ref. 1, p. 162, we have $\overline{n_c(t_1)n_s(t_2)} = -\overline{n_s(t_1)n_c(t_2)}$ so the cross terms of (3.10) add up to zero. Since the noise is stationary, the autocorrelation depends only on the difference $\tau = t_1 - t_2$. Denoting autocorrelation by $R(\tau)$, we find the autocorrelation of n' to be

$$R_{n'}(\tau) = \frac{1}{V_s^2}\left[R_{nc}(\tau)\cos^2\theta_o + R_{ns}(\tau)\sin^2\theta_o\right] \quad (3.11)$$

Taking Fourier transforms, the spectrum of n' is

$$\Phi_{n'}(f) = \frac{1}{V_s^2}\left[\Phi_{nc}(f)\cos^2\theta_o + \Phi_{ns}(f)\sin^2\theta_o\right] \quad (3.12)$$

However, $\Phi_{ns}(f) = \Phi_{nc}(f) = \Phi_n(f_i - f) + \Phi_n(f_i + f)$, where $\Phi_n(f)$ is the one-sided spectral density of the bandpass input noise $n(t)$. (Note that f is nonnegative because we are considering one-sided spectra. Also, $f_i = \omega_i/2\pi$.)

The spectrum of n' reduces to

$$\Phi_{n'}(f) = \frac{1}{V_s^2}\left[\Phi_n(f_i - f) + \Phi_n(f_i + f)\right] \quad (3.13)$$

---

*Loop bandwidth is defined in (3.17).

For the special case of white noise, $\Phi_n(f) = N_o$ V^2/Hz, so the spectrum of n' becomes

$$\Phi_{n'}(f) = \frac{2N_o}{V_s^2} \qquad (3.13a)$$

The only assumptions underlying these results are that the input noise is bandpass, stationary, gaussian noise and that the VCO phase is time invariant, though otherwise arbitrary. Under these assumptions, n' must be gaussian. So far there has been no linearizing approximation with respect to noise.

(Noise output of the phase detector is $K_d n'(t)$. Such an output could be caused by the additive noise, as described, or it could be caused by an input phase disturbance $\sin \theta_{ni}(t) = n'(t)$. If θ_{ni} is small enough, the sinusoidal nonlinearity can be neglected and the variance of the fictitious input phase disturbance is $\overline{\theta_{ni}^2} = \sigma_{n'}^2 = \sigma_n^2/V_s^2$. Input signal to noise ratio is $SNR_i = V_s^2/2\sigma_n^2$, so the input phase variance is approximated by $\overline{\theta_{ni}^2} = 1/2SNR_i$. That is the jitter to be expected if one were to measure the phase difference between a clean signal and one corrupted by noise, under conditions of large signal-to-noise ratio. This relation is used later in establishing a definition for signal-to-noise ratio of the phaselock loop.)

(In light of the duality between n' and θ_{ni}, it is sometimes useful to consider n' to be an angle disturbance, with units of radians—a dimensionless quantity. Then the spectral density $\Phi_{n'}$ can be considered to have units of rad^2/Hz.)

The assumption of time-invariant θ_o has implied an open loop; if the loop were closed, the noise would frequency modulate the VCO and θ_o would fluctuate in random fashion. Our eventual goal in this analysis is to determine the properties of the fluctuations of θ_o.

The phase detector is nonlinear, so the fed-back fluctuations intermodulate with the incoming signal plus noise. Completely general analysis is blocked by this nonlinearity; all useful results require simplifying approximations.

The most common simplification is to assume noise is small enough that $(\theta_i - \theta_o)$ remains small and that the PD can be regarded as linear. Under these conditions, the intermodulation products may be neglected and a linearized phaselock loop, with simple, additive noise $n'(t)$, may be considered, as shown in Figure 3.2.

From Figure 3.2 it becomes evident that $n'(t)$ is simply additive to the input-signal phase θ_i. In Chapter 2 a transfer function $H(j\omega)$ is derived, which relates θ_o to θ_i; the same transfer function relates θ_o to $n'(t)$. Spectral density Φ_{no} of the VCO phase is related to the spectrum of n' by the

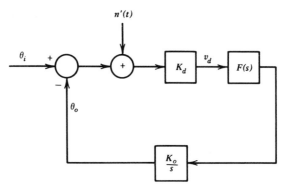

Figure 3.2 Block diagram of linearized loop.

relation

$$\Phi_{no}(f) = \Phi_{n'}(f)|H(j\omega)|^2$$

$$= \frac{1}{V_s^2}[\Phi_n(f_1-f) + \Phi_n(f_i+f)]|H(j\omega)|^2 \qquad (3.14)$$

Variance of the output phase is

$$\overline{\theta_{no}^2} = \int_0^\infty \Phi_{n'}(f)|H(j\omega)|^2\,df \qquad \text{rad}^2 \qquad (3.15)$$

where, as usual, $\omega = 2\pi f$.

In general, the integral of (3.15) is cumbersome to evaluate. However, it simplifies radically in the important special case of white input noise. If $\Phi_n(f) = N_o$ (V²/Hz) for all frequencies of interest, then (3.15) becomes

$$\overline{\theta_{no}^2} = \frac{2N_o}{V_s^2}\int_0^\infty |H(j2\pi f)|^2\,df \qquad (3.16)$$

The integral of (3.16) defines the *noise bandwidth* of the loop and commonly is given the symbol B_L.

$$\boxed{B_L = \int_0^\infty |H(j2\pi f)|^2\,df \qquad \text{Hz}} \qquad (3.17)$$

Therefore, if the input noise applied to the loop is white, the phase variance is given by the very simple formula

$$\boxed{\overline{\theta_{no}^2} = \frac{2N_o B_L}{V_s^2} = \frac{W_i B_L}{P_s} \quad \text{rad}^2} \tag{3.18}$$

where P_s is the signal power in watts and W_i is the noise power spectral density in W/Hz.

The integral of (3.17) has been evaluated explicitly for several common varieties of loop transfer functions and the resulting expressions for noise bandwidth are shown in Table 3.1 (refer to Chapter 2 for definitions of notation). Notice that B_L has dimensions of Hertz, despite the fact that K, a_2, and ω_n are given in radians per second.

The central role of the loop gain K in establishing B_L is evident from the table. It is intriguing to see that addition of a simple lag filter to a first-order loop does not affect the noise bandwidth; this is one reason for regarding the second-order loop with lag filter as a modified first-order loop rather than a genuine second-order loop.

Noise bandwidth of the popular high-gain, second-order loop is plotted against damping in Figure 3.3. Minimum noise bandwidth is achieved for $\zeta = \frac{1}{2}$. Noise bandwidth does not exceed the minimum by more than 25% (1 dB additional jitter) for any damping between 0.25 and 1.0.

Table 3.1   Noise Bandwidths of Common Loops

Loop Description	Noise Bandwidth, B_L(Hz)
First order	$\frac{1}{4}K$
Second order:	
Simple lag filter	$\frac{1}{4}K$
Passive lag–lead filter	$\frac{1}{4}K\dfrac{K + a_2 + 1/\tau_1}{K + 1/\tau_1}$
	$\simeq \frac{1}{4}K\left(1 + \dfrac{a_2}{K}\right)$ if $K \gg 1/\tau_1$ and $a_2 \gg 1/\tau_1$
Active lag–lead filter	$\frac{1}{4}K\left(1 + \dfrac{a_2}{K}\right)$ or $\frac{1}{2}\omega_n\left(\zeta + \dfrac{1}{4\zeta}\right)$
Third order: Two zeros; all open-loop poles at origin	$\frac{1}{4}K\dfrac{a_2 K + a_2^2 - a_3}{a_2 K - a_3}$

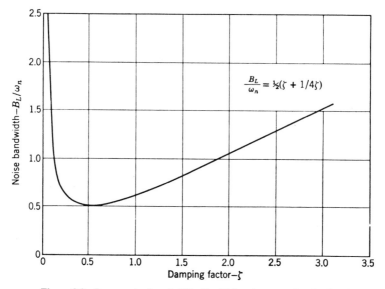

Figure 3.3 Loop noise bandwidth (for high-gain, second-order loop).

Signal-to-noise ratio (SNR) is a useful engineering concept and it is often helpful to define one for the phaselock loop. Definition of input signal-to-noise ratio SNR_i is straightforward; it is merely the ratio of input signal power to input noise power.

By contrast, there is no "signal" within the PLL; normal tracking is about a null. Also, loop "noise" is a function of the point at which the measurements are performed—there is no unique definition. As a result, loop signal-to-noise ratio SNR_L must be defined arbitrarily and is a fictitious quantity without firm physical meaning.

In this book SNR_L is defined by analogy among phase jitters. The input phase jitter, for large SNR_i, was found to be

$$\overline{\theta_{ni}^2} = \frac{1}{2SNR_i} \qquad (3.19)$$

By analogy, we define SNR_L arbitrarily from

$$\overline{\theta_{no}^2} = \frac{1}{2SNR_L} \qquad (3.20)$$

Then, using (3.18), we arrive at

$$\boxed{SNR_L = \frac{P_s}{2B_L W_i} = \frac{V_s^2/2}{2B_L N_o}} \qquad (3.21)$$

Equation 3.21 is taken as the definition of loop signal-to-noise ratio for all values of SNR_L, large or small. However, (3.18) and (3.20), which were used in generating (3.21), are valid only for large SNR_L. Nonlinear operation (at small SNR_L) is considered later in the chapter.

[Another common definition of loop signal-to-noise ratio is $P_s/B_L W_i$; it and (3.21) are equally valid and equally arbitrary. Care must be taken in reading phaselock literature to ascertain which definition is used by the author.]

Despite the fictitious nature of SNR_L, it is possible to endow it with a useful conceptual meaning. Let us consider that the loop acts as a band-pass filter around the received signal. The filter is centered at the frequency of the signal and has a noise bandwidth of B_L on each side of center for a total equivalent input bandwidth of $2B_L$. Thus, for white noise of spectral density W_i, the total noise power that enters the loop is $2B_L W_i$ watts. The ratio of signal power to this value of noise power is our definition (3.21) of SNR_L.

Like all bandwidths in this book, B_L is a one-sided bandwidth (see Appendix A). One might reasonably and properly call $2B_L$ the *double sideband noise bandwidth* of the loop; $2B_L$ is one-sided. The distressing, though common, practice of describing $2B_L$ as the two-sided noise band-width is improper, especially in conjunction with one-sided noise spectra.

3.2 NONLINEAR OPERATION

Observed Behavior

When a PLL is monitored in the laboratory, the phase jitter of the VCO is observed to be more than is predicted by (3.18) and (3.20), as SNR_L is reduced below about 4 dB. See curve *a* in Figure 3.4. The discrepancy should cause no surprise, since the linear analysis was based on an assumption of small phase error in the loop, but the actual error at low SNR_L is not small. The linear analysis fails when its underlying assumption is violated.

Another phenomenon appears at low SNR_L; the oscillator phase occasionally slips one or more cycles as compared to the signal. A large noise event, in effect, knocks the loop temporarily out of lock and tracking returns to equilibrium n cycles away from its original condition ($n = \pm 1, \pm 2$, etc.). Frequency of slipping is a very steep function of SNR_L, as shown in Figure 3.5. Cycle slips are particularly destructive to operations in which every cycle counts, such as Doppler velocity measurements or recovery of digital clock timing (Chapter 11). Slips are also important to the understanding of phaselocked FM demodulators (Chapter 9).

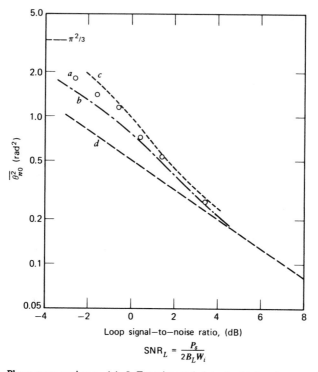

Figure 3.4 Phase-error variance. (*a*) ○ Experimental data for high-gain, second-order loop; $\zeta = 0.707$ (Ref. 7). (*b*) Exact nonlinear analysis for first-order loop (Ref. 5). (*c*) Approximate nonlinear analysis for second-order loop; $\zeta = 0.707$ (data from Refs. 12 and 13). (*d*) Linear approximation; $\overline{\theta_{n_o}^2} = 1/(2\mathrm{SNR}_L)$.

A third phenomenon emerges if SNR_L is reduced sufficiently; the loop drops out of lock and stays out. Control of the VCO is lost; its frequency wanders off from the signal frequency. Although both phenomena have often been lumped together under the name *loss of lock*, drop lock is qualitatively distinct from repeated cycle slips.

The drop-lock SNR_L typically is in the vicinity of 0 dB, although extreme care with loop components may extend the dropout point 1 to 2 dB lower. An observer gets the impression that the loop at low SNR_L is staggering (because of repeated cycle slips) and eventually everything seems to collapse as lock is lost completely. Reacquisition of lock is nearly impossible after dropout, unless SNR_L is raised substantially (to about 3 to 6 dB).

Experience of the drop-lock phenomenon led to the concept of a *noise threshold* of the PLL; that is, the loop falls out of lock if SNR_L is below the

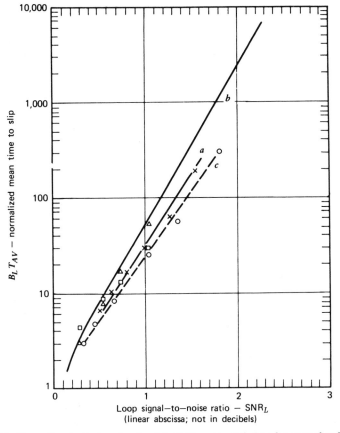

Figure 3.5 Mean time to first cycle slip. (*a*) × Experimental data for second-order loop; $\zeta = 0.707$ (Ref. 7). (*b*) Exact result for first-order loop (3.25). (*c*) ○ Simulations for second-order loop; $\zeta = 0.707$ (Ref. 17). (*d*) △ Simulations for second-order loop; $\zeta = 1.4$ (Ref. 16). (*e*) □ Simulations for second-order loop; $\zeta = 0.35$ (Ref. 16).

"threshold" level. The idea of threshold was given support at one time by *quasi-linear* analysis.[2,3]

Later, it gradually became apparent that well-built loops could hold lock below the analytical threshold, whereupon it was realized that the predicted threshold was a feature of the approximations in the analysis and not of a real PLL. Despite the incorrect prediction of threshold, the quasi-linear method—and its modifications[4]—provides a simple method of adequate engineering accuracy for predicting phase jitter in a practical operating range of SNR_L.

Presently accepted nonlinear analyses do not reveal a noise threshold (see next section). Current opinion holds that drop lock arises from a complicated nonlinear interaction between the noise-caused phase jitter and small biases and DC offsets arising from imperfections of the loop components, especially the phase detector (see Chapter 6). The imperfections are circuit specific and are usually unpredictable,. even after the circuit is built. In general, drop lock would be very difficult to analyze and any analysis would be difficult to apply.

Analytical difficulties aside, this viewpoint sees drop lock as a technological problem and not inherent to the PLL, as such. If the viewpoint is correct, drop lock can, in principle, be pushed to lower signal-to-noise ratios by improvements in the loop components.

By contrast, analyses described in the sequel predict cycle slipping characteristics very well. Moreover, the cycle-slip predictions are for ideal loops, so no relief can be obtained from improvements in loop components.

Nonlinear Analysis of Phase Error

In a linear system, a gaussian input gives rise to a gaussian output. Therefore, our earlier assumptions of linear operation of the loop and of gaussian input noise imply that the VCO phase jitter would be gaussian. A gaussian process is completely defined by its autocorrelation function or, equivalently, its spectral density as derived in (3.14). Variance is readily found from either one.

Response of a nonlinear system to a gaussian stimulus is generally nongaussian and the second-order statistics do not define the process completely. Nonlinear analysis of a PLL has been concerned with deriving the nongaussian probability density function (pdf) of the phase error, computing the phase variance from the pdf, and investigating the statistics of cycle slipping.

The analytical simplicity of transfer functions is lost in a nonlinear system. Analysis of a nonlinear system is much more difficult and demands a higher level of mathematical sophistication than linear analysis. The treatment here presents a summary of the results of the various nonlinear analyses, a treatment that is more than sufficient for most engineering purposes. (In fact, linear analysis will suffice for the great bulk of engineering design problems.) References are provided for those who are interested in the detailed mathematics.

Viterbi's exact analysis[5] of the first-order loop has provided much insight and many useful tools for understanding nonlinear operation. First, one must recognize that cycle slips cause the phase error $(\theta_i - \theta_o)$ to be a growing quantity and ultimately unbounded; that is, phase error, as it has been defined so far, is nonstationary, so our well-honed tools of stationary analysis are not applicable directly.

To avoid this problem, we define a new phase variable

$$\phi = (\theta_i - \theta_o) \text{ modulo } 2\pi \qquad (3.22)$$

so that, although $(\theta_i - \theta_o)$ can take on any value from $-\infty$ to $+\infty$, the value of ϕ is bounded since (3.22) means $(\theta_i - \theta_o) = \phi + 2n\pi$, where n is chosen to cause ϕ to lie in the interval $(-\pi, \pi)$.

This definition of ϕ implies that all cycles of a sine wave look alike and cannot readily be distinguished from one another. Cycle slips are neglected by this definition and they must be treated separately. Most laboratory instruments operate modulo 2π and therefore yield ϕ rather than $\theta_i - \theta_o$; the concept agrees well with normal practice, despite any initial impressions of peculiarity.

It turns out that ϕ is stationary (in the steady state, after any transients have died out), which allows us to apply stationary statistics. Let us denote the probability-density function of ϕ as $p(\phi)$; this is found as the steady-state solution of a nonlinear, stochastic partial-differential equation known as the Fokker-Planck equation. Bypassing details[5] we have

$$p(\phi) = \frac{\exp(\rho \cos\phi)}{2\pi I_o(\rho)} \qquad |\phi| \leqslant \pi \qquad (3.23)$$

where $\rho = 2\text{SNR}_L$ and $I_o(\rho)$ is the modified Bessel function of the first kind and zero order.[6]

(Equation 3.23 is valid only if the static phase error $\overline{\phi} = 0$. See Chapter 4 for an explanation of static phase error. See Ref. 12 for $p(\phi)$ if $\overline{\phi} \neq 0$.)

The density (3.23) approaches gaussian for large SNR_L, thereby agreeing with the linear analysis. At very small SNR_L, $p(\phi)$ approaches a uniform density over $(-\pi, \pi)$, which is characteristic of the phase of random noise.

Variance of the phase error can be found by numerical evaluation of

$$\overline{\phi^2} = \int_{-\pi}^{\pi} \phi^2 p(\phi)\, d\phi \qquad (3.24)$$

The result, phase variance reduced modulo 2π, is plotted in curve b of Figure 3.4. The exact variance agrees with the linear analysis for large SNR_L and approaches $\pi^2/3$ for very small SNR_L. (The variance of a random variable, uniformly distributed over $(-\pi, \pi)$, is $\pi^2/3$.)

Knowledge of variance is useful but is insufficient by itself because cycle slipping is evaded in its computation. Statistics of cycle slipping are an important attribute of PLL operation at low SNR_L, as is the phase variance.

By means of manipulations on the Fokker-Planck equations, Viterbi derives an expression for the average time T_{AV} between cycle slips. From a condition of zero phase error, T_{AV} is the average time required for the loop phase error to reach $\pm 2\pi$ for the first time. If slipping occurs primarily as single, isolated slips, then the frequency of cycle slips is $1/T_{AV}$. If slips cluster—as may happen in a second- or higher-order loop— then T_{AV} and the slip rate are not quite so simply related.

For the first-order loop with zero static phase error

$$T_{AV} = \frac{\pi^2 \rho I_o^2(\rho)}{2B_L} \tag{3.25}$$

which is approximated for large ρ by

$$T_{AV} \simeq \frac{\pi}{4B_L} \exp(2\rho) \tag{3.26}$$

A plot of T_{AV} from (3.25) is shown as curve b in Figure 3.5; the straightness shows that (3.26) is acceptable for all practical SNR_L.

In addition, time between slips is exponentially distributed; that is, the probability that the loop will slip within T sec, starting from zero error, is

$$P(T) = 1 - \exp(-T/T_{AV}) \tag{3.27}$$

This distribution is well confirmed by computer simulations and laboratory measurements on both first- and second-order loops.

Viterbi's results apply exactly to a first-order loop with sinusoidal phase detector and additive, white, gaussian noise. The first-order pdf (3.23) and variance also apply,[7,8] without modification, if a simple lag filter of the form $F(s) = (s\tau + 1)^{-1}$ is inserted into the loop, provided static phase error is zero.

Fokker-Planck equations can be written for the true second-order loop, but explicit, exact solutions have been unattainable. The second-order loop is technologically the most important configuration, so there is a strong motivation to determine its statistics for low SNR_L. Experimental measurements of $p(\phi)$ and $\overline{\phi^2}$ are reported in Ref. 7. The measured variance is shown as experimental points (curve a) of Figure 3.4. We can draw two significant conclusions:

1. Exact nonlinear analysis of the first-order loop is in close agreement with measured performance of the second-order loop for SNR_L in excess of 0 dB, that is, for any useful value of SNR_L.

2. Approximate linear analysis yields good accuracy if SNR_L is in excess of 5–6 dB.

Because of the practical importance of a second-order PLL, numerous approximate analyses have been devised.[7-14] These analyses generally involve clever assumptions and heroic mathematics.

Among the several methods, Lindsey's approach[12,13] starts with the Fokker-Planck equation, so it yields approximations to the pdf and slip statistics as well as phase variance. He provides substantial detail in the form of charts and formulas. His prediction of variance is shown as curve c in Figure 3.4. The approximation is clearly very close to the measured results, albeit slightly pessimistic. Similar agreement is found in comparing the measured pdf against the predicted value.

Lindsey's analysis predicts that the phase variance will have a weak inverse dependence on damping factor ζ; that is, jitter is slightly worse for small damping, given the same loop bandwidth B_L and same SNR_L. The prediction is borne out by simulation results.[15] However, if SNR_L exceeds unity—as it must in a useful loop—then the spread between light damping ($\zeta = 0.35$) and a first-order loop ($\zeta = \infty$) is small and may be neglected for most purposes.

Several investigators have studied slipping in a second-order PLL by means of computer simulation and by measurements on physical loops in the laboratory. A summary of their published results is given in Figure 3.5.

It is evident that slipping becomes somewhat worse if damping is small. The first-order loop has infinite damping so its slip time T_{AV} is greater than that of any second-order loop of the same noise bandwidth.

(The experimental curves suffer from statistical fluctuations due to a finite number of samples in the measurements and from *ad hoc* redefinitions of the meaning of "slip."[16] Some caution must be exercised in applying the data.)

Predictions of T_{AV} for a second-order loop are given by formulas developed (at great effort) by Lindsey[12,13] and Tausworthe,[17,18] predictions in fairly close agreement with the experiments. Since the formulas are cumbersome and since their derivation necessarily involved approximations, the practicing engineer will usually find the curves of Figure 3.5 to be a more convenient guide to slip behavior.

Inspection of the data points of Figure 3.5 shows a reasonably good fit to a straight line when they are plotted on a logarithmic ordinate. This means that T_{AV} is approximately exponentially dependent on SNR_L. Curves b and c of the figure represent boundary limits that encompass all the configurations that have been explored. Curve b is the exact result for a first-order loop and is described by (3.25) and (3.26). Curve c is from simulation of a second-order loop with damping 0.707; its level is thought to be somewhat pessimistic. The points of c are well fitted by an empirical relation (valid only for zero static-phase error)

$$B_L T_{AV} = \exp(\pi SNR_L) \tag{3.28}$$

It appears to be reasonable to use (3.26) and (3.28) as upper and lower bounds on T_{AV}.

Some numbers are of interest. If we let $B_L = 20$ Hz, at $SNR_L = 1$ (0 dB) (3.28) predicts $T_{AV} = 1.16$ sec, which would be very poor performance. Use of (3.26) predicts $T_{AV} = 2.1$ sec, which is also very poor.

Now let us consider $SNR_L = 10$; the lower bound on T_{AV} is predicted to be 2.2×10^{12} sec or about 70,000 years. (We assume, without any experimental verification, that the exponential relation can be extrapolated to large SNR.)

Miscellaneous Features

The most important descriptors of nonlinear operation are the pdf and variance of the phase error, reduced modulo 2π, and the slip statistics. This information has been presented for a loop that is unstressed by any other phase error, such as may be caused by steady phase error or angle modulation. Chapter 4 discusses the origins of such phase errors and how they may be reduced.

The presence of a static phase error causes the phase jitter to increase and the presence of noise causes any static phase error to increase from its no-noise level. An appealing physical insight is provided by the approximate analysis of Blanchard.[4] Other analyses are given in Refs. 12, 13, and 19.

As might well be imagined, presence of a phase error increases the propensity to cycle slipping. The effect of a static phase error is expounded in Refs. 12, 13, 16–18, 20, and 25. Virtually no information exists on slipping in the presence of time-varying phase error.

Observations of cycle slips have revealed that the slips of a first-order loop are almost always single, isolated events. Therefore, the rate of cycle slipping r_s is simply the reciprocal of the mean time to first slip: $r_s = 1/T_{AV}$.

In a second-order loop, the slips tend to bunch in bursts; the average number of slips per burst increases with worsening SNR_L. See Refs. 15, 16, 22, and 23. No good physical explanation has been advanced for this behavior, although explanation of a related phenomenon is attempted in Chapter 4. The slip rate $r_s \neq 1/T_{AV}$ in a second-order loop.

All of the foregoing nonlinear analysis, simulation, and measurement applies only to white noise. In practice, "white" means that the bandwidth of the input noise is large compared to the noise bandwidth $2B_L$ of the PLL. The Fokker-Planck approach is inapplicable to analysis with narrowband noise. In consequence, we are unable to predict cycle slipping for narrowband noise inputs. In Chapter 9, it is seen how this restriction has impeded understanding of phaselocked FM demodulators.

Hess[24] has devised an approximate analysis of cycle slipping in a first-order loop exposed to bandlimited noise. His formulas are confirmed by measurements of cycle slipping on laboratory PLLs.

All of the foregoing has assumed that gaussian noise was applied to the PLL; different noise statistics require modified analysis. A limiter is often used in front of a phase detector, making the noise statistics decidedly nongaussian. Discussion of the effect of a limiter in the nonlinear region of loop operation may be found in Refs. 13, 21, and 24. Effect of a limiter in the linear region is examined in Chapter 6.

Finally, the known information on nonlinear operation is confined to first- and second-order loops; there is almost nothing published on higher-order loops. Since the third-order loop is of some practical importance, this lack of data is a barrier to fully understood design. The only present expedient is to assume that a third-order loop behaves much the same as a second-order loop of the same noise bandwidth.

REFERENCES

1. W. B. Davenport and W. L. Root, *Random Signals and Noise*, New York, McGraw-Hill, 1958.

2. J. A. Develet, Jr., "A Threshold Criterion for Phase-Lock Demodulator," *Proc. IEEE*, Vol. 51, pp. 349–356, February 1963.

3. J. A. Develet, Jr., "An Analytic Approximation of Phase-Lock Receiver Threshold," *IEEE Trans.*, SET-9, pp. 9–11, March 1963.

4. A. Blanchard, "Phase-Locked Loop Behavior Near Threshold," *IEEE Trans.*, AES-12, pp. 628–638, September 1976; corrections: *AES-12*, p. 823, November, 1976.

5. A. J. Viterbi, *Principles of Coherent Communications*, New York, McGraw-Hill, 1966.

6. M. Abramowitz and I. A. Stegun, *Handbook of Mathematical Functions*, National Bureau of Standards Publication 55, U. S. Government Printing Office, Washington, 1964, Chap. 9.

7. F. J. Charles and W. C. Lindsey, "Some Analytical and Experimental Phaselocked Loop Results for Low Signal-to-Noise Ratios," *Proc. IEEE*, Vol. 54, pp. 1152–1166, September 1966.

8. A. Blanchard, *Phase Locked Loops*, Wiley, New York, 1976, Chap. 12.

9. H. L. Van Trees, "Functional Techniques for the Analysis of the Nonlinear Behavior of Phase-Locked Loops," *Proc. IEEE*, Vol. 52, pp. 894–911, August 1964.

10. R. C. Tausworthe, *Theory and Practical Design of Phase-Locked Receivers*, Vol. I, Technical Report No. 32-819, Jet Propulsion Laboratory, Pasadena, CA, February 1966.

11. J. K. Holmes, "On a Solution to the Second-Order Phase-Locked Loop," *IEEE Trans.*, *COM-18*, pp. 119–126, April 1970.

12. W. C. Lindsey, *Synchronization Systems in Communication and Control*, Prentice-Hall, Englewood Cliffs, NJ, 1972.

13. W. C. Lindsey and M. K. Simon, *Telecommunication Systems Engineering*, Prentice-Hall, Englewood Cliffs, NJ, 1973, Chap. 2.

14. H. Meyr, "Nonlinear Analysis of Correlative Tracking Systems Using Renewal Process Theory," *IEEE Trans.*, *COM-23*, pp. 192–203, February 1975.

15. J. R. Rowbotham and R. W. Sanneman, "Random Characteristics of the Type II Phase-Locked Loop," *IEEE Trans.*, *AES-3*, pp. 604–612, July 1967.

16. R. W. Sanneman and J. R. Rowbotham, "Unlock Characteristics of the Optimum Type II Phase-Locked Loop," *IEEE Trans.*, *ANE-11*, pp. 15–24, March 1964.

17. R. C. Tausworthe, "Cycle Slipping in Phase-Locked Loops," *IEEE Trans.*, *COM-15*, pp. 417–421, June 1967.

18. R. C. Tausworthe, "Simplified Formula for Mean Cycle-Slip Time of Phase-Locked Loops with Steady-State Phase Error," *IEEE Trans.*, *COM-20*, pp. 331–337, June 1972.

19. W. C. Lindsey and M. K. Simon, "The Effect of Loop Stress on the Performance of Phase-Coherent Communications," *JPL SPS 37-56*, Vol. III, pp. 104–118, Jet Propulsion Laboratory, Pasadena, CA, April 1969.

20. E. A. Bozzoni, G. Marchetti, U. Mengali, and F. Russo, "An Extension of Viterbi's Analysis of Cycle Slipping in a First-Order Phase-Locked Loop," *IEEE Trans.*, *AES-6*, pp. 484–490, July 1970.

21. R. C. Tausworthe, "Cycle-Slipping in the Second-Order Loop with Limiting," *JPL SPS 37-56*, Vol. II, pp. 94–97, Jet Propulsion Laboratory, Pasadena, CA, March 1969.

22. S. C. Gupta, J. W. Bayless, and D. R. Hummels, "Threshold Investigation of Phase-Locked Discriminators," *IEEE Trans.*, *AES-4*, pp. 855–862, November 1968.

23. B. M. Smith, "The Phase-Lock Loop with Filter: Frequency of Skipping Cycles," *Proc. IEEE*, Vol. 54, p. 296, February 1966.

24. D. T. Hess, "Cycle-Slipping in a First-Order Phase-Locked Loop," *IEEE Trans.*, *COM-16*, pp. 255–260, April 1968.

25. J. K. Holmes, "First Slip Times Versus Static Phase Error Offset for the First- and Passive Second-Order Phase-Locked Loop," *IEEE Trans.*, *COM-19*, p. 234, April 1971.

Chapter Four

Tracking

4.1 LINEAR TRACKING

To study tracking we examine the phase error θ_e that results from a specified input θ_i. A small phase error is usually desired and is considered to be the criterion of good tracking performance. If the error should become so large that the VCO skips cycles, the loop is considered to have lost lock, even if only momentarily. The problems of unlock behavior are considered in a later section. Here the concern is with tracking in a locked loop with phase error small enough to justify an assumption of linearity.

Steady-State Errors

Phase error (in the frequency domain) is given by (2.7) as

$$\theta_e(s) = \frac{s\theta_i(s)}{s + K_o K_d F(s)} \tag{4.1}$$

The simplest to analyze are the steady-state errors remaining after any transients have died away. These errors are readily evaluated by means of the final value theorem of Laplace transforms, which states

$$\lim_{t \to \infty} y(t) = \lim_{s \to 0} sY(s) \tag{4.2}$$

That is to say, the steady-state value of a function in the time domain is readily determined from inspection of its transform in the frequency domain.

Application of the final value theorem to the phase-error equation yields

$$\lim_{t \to \infty} \theta_e(t) = \lim_{s \to 0} \frac{s^2 \theta_i(s)}{s + K_o K_d F(s)} \tag{4.3}$$

43

As a first example, let us consider the steady-state error resulting from a step change of input phase of magnitude $\Delta\theta$. The Laplace transform of the input is therefore $\theta_i(s) = \Delta\theta/s$, which may be substituted into (4.3) to give

$$\lim_{t\to\infty} \theta_e(t) = \lim_{s\to 0} \frac{s\,\Delta\theta}{s + K_o K_d F(s)} = 0$$

[provided that $F(0) > 0$]. In other words, the loop will eventually track out any change of input phase; there is no steady-state error resulting from a step change of phase.

For another example, let us examine the steady-state error resulting from a step change of frequency of magnitude $\Delta\omega$. Input phase is a ramp, $\theta_i(t) = \Delta\omega t$, so $\theta_i(s) = \Delta\omega/s^2$. Substitution of this value of θ_i into (4.3) results in

$$\theta_v = \lim_{t\to\infty} \theta_e(t) = \lim_{s\to 0} \frac{\Delta\omega}{s + K_o K_d F(s)} = \frac{\Delta\omega}{K_o K_d F(0)} \qquad (4.4)$$

The product $K_o K_d F(0)$ is often called the *velocity constant* or *DC loop gain* and is denoted by the symbol K_v. Those familiar with servos will recognize it as the velocity-error coefficient. Note that K_v has the dimensions of frequency.

Incoming signal frequency almost never agrees exactly with the free-running (zero control voltage) frequency of the VCO. As a rule there is a frequency difference $\Delta\omega$ between the two. The frequency difference may be due to an actual difference between the transmitter and receiver or it may be due to a Doppler shift. In either case the resulting phase error is often called the *velocity error*, *loop stress*, or *static phase error* and is given by

$$\boxed{\theta_v = \frac{\Delta\omega}{K_v}} \qquad (4.4a)$$

A heuristic derivation of (4.4) provides better physical insight. The control voltage needed to retune the VCO by an amount $\Delta\omega$ is $\Delta\omega/K_o$. In the steady state, the control voltage $v_c = v_d F(0)$, where v_d is the DC output of the phase detector. But phase detector output is produced by a phase error $\theta_e = v_d/K_d$. Therefore, to produce the necessary control voltage requires the phase error $\theta_e = \Delta\omega/K_o K_d F(0)$ as in (4.4).

Let us evaluate K_v for the second-order loop. Two types of loop filter are considered in Chapter 2: a passive filter and an active filter. For the passive filter $F(0) = 1$, whereas for the active filter $F(0) = A$, where A is the

DC gain of an operational amplifier. Assuming that $K_o K_d$ is the same in both cases, we see that K_v will be much larger and θ_v much smaller if an active filter is used. (Voltage gains of 10^2 to 10^7 are typical.) As a practical matter, it is not difficult, in most cases, to make A large enough that θ_v is no more than a few degrees for the maximum frequency difference encountered.

We can now see the reason for the predominance of second-order loops. Since signal frequency almost invariably is unequal to the free-running frequency of the VCO, a nonzero control voltage is needed to tune the VCO and hold the loop in lock. The integrator in the loop filter of a second-order loop generates the necessary control voltage while still permitting small phase error. Moreover, the large integrator gain can be obtained simultaneously with small noise bandwidth, an impossible combination in a first-order loop.

Next, let us suppose that the input frequency is linearly changing with time at a rate of $\Delta\dot{\omega}$ rad/sec^2; that is, $\theta_i(t) = \frac{1}{2}\Delta\dot{\omega}t^2$. Such input behavior might arise from accelerated motion between transmitter and receiver, from change of Doppler frequency during an overhead pass of a satellite, or from sweep-frequency modulation. Transformed phase is $\theta_i(s) = \Delta\dot{\omega}/s^3$, and it can be shown that phase error will grow without bound if K_v is finite.

However, let us suppose that an active loop filter is used and that K_v is large enough that static phase error (4.4a) is negligible for the largest frequency excursions to be accommodated. Under this assumption an ideal integrator, with infinite K_v, is a good approximation to the actual filter. Then from (2.12) the phase error in a second-order loop may be written as

$$\theta_e(s) = \frac{s^2\theta_i(s)}{s^2 + 2\zeta\omega_n s + \omega_n^2} \tag{4.5}$$

By use of (4.2) this leads to the *acceleration error* (sometimes called *dynamic tracking error* or *dynamic lag*)

$$\theta_a = \lim_{t\to\infty} \theta_e(t) = \lim_{s\to 0} \frac{\Delta\dot{\omega}}{s^2 + 2\zeta\omega_n s + \omega_n^2} \tag{4.6}$$

$$\boxed{\theta_a = \frac{\Delta\dot{\omega}}{\omega_n^2} \quad \text{rad}} \tag{4.7}$$

It is possible to obtain (4.7) from physical considerations. We apply a DC voltage v_d to the integrator of the loop filter. Integrator output is

$v_c(t) = v_c(0) + v_d t / \tau_1$, so the rate of change of VCO frequency is $\Delta\dot\omega = K_o v_d / \tau_1$. The DC voltage v_d must be generated by a phase error $\theta_e = v_d / K_d$, which, when substituted into the expression for frequency rate, gives $\Delta\dot\omega = K_o K_d \theta_e / \tau_1$. But, from (2.11), $K_o K_d / \tau_1 = \omega_n^2$, whereupon (4.7) follows.

It is sometimes necessary to track an accelerating phase without incurring steady-state tracking error. Let us determine the form of $F(s)$ needed to reduce θ_a to zero. The expression for final value acceleration error is

$$\theta_a = \lim_{s \to 0} \frac{\Delta\dot\omega}{s[s + K_o K_d F(s)]} \tag{4.8}$$

For θ_a to be zero, it is necessary that $F(s)$ have the form $Y(s)/s^2$, where $Y(0) \neq 0$. The factor $1/s^2$ implies that the loop filter must contain two cascaded integrators. Closed-loop response then has a polynomial of third degree in its denominator, and the loop is of third order. Because of this property of eliminating the steady-state acceleration error, a third-order loop is sometimes very useful in tracking satellites and missiles.[6]

In a second-order loop, it is necessary to employ a large natural frequency, and therefore, large noise bandwidth, to handle a rapidly changing input frequency. By going to third order, the frequency rate can be accommodated in a loop with small noise bandwidth.

One other steady-state error is ever present. It is caused by unwanted DC offsets in the active filter and in the phase detector. The loop acts to produce a DC null, including the effect of offset, so the phase error needed to counteract the offset is simply the offset voltage divided by K_d, the PD gain factor. Further discussion of offsets is found in Chapter 6.

Transient Response

Besides steady-state behavior, it is often necessary to determine the transient phase error caused by particular inputs. The signal phases considered in the last section are:

1. A step of phase, $\Delta\theta$ rad.
2. A step of frequency (phase ramp), $\Delta\omega$ rad/sec.
3. A step of acceleration (frequency ramp), $\Delta\dot\omega$ rad/sec^2.

For these inputs, the L-transformed input phase is $\Delta\theta/s, \Delta\omega/s^2$, and $\Delta\dot\omega/s^3$ respectively. To compute phase errors each input is substituted into (4.1) and inverse L-transforms are then computed (or looked up in tables) to determine time response.

In a first-order loop, the resulting transient phase errors are simple exponentials:

$$\Delta\theta e^{-Kt} \qquad\qquad \text{(phase step)}$$

$$\frac{\Delta\omega}{K}(1-e^{-Kt}) \qquad\qquad \text{(frequency step)}$$

$$\frac{\Delta\dot{\omega}}{K^2}(Kt+e^{-Kt}-1) \qquad \text{(frequency ramp)}$$

Note that the frequency-ramp response increases with time and that linear bounds are eventually passed. The analyses in this section are all predicated on a linear approximation and all fail if the loop is driven into a nonlinear region.

The analytic expressions for phase error of a high-gain, second-order loop are given in Table 4.1 and are plotted in Figures 4.1 to 4.3. The plots are from Hoffman's monograph,[1] which contains many additional useful plots of similar nature.

A third-order loop can be treated in the same manner, but the published results[7-9] are few and are widely scattered. The reason lies partly in the far-greater popularity of the second-order loop, but also in the extra complexity of third order. There are three loop parameters in a third-order loop, so many pages of figures would be needed to present the same kind of data as in, for example, Figure 4.2.

As a rough rule of thumb, one can assume that the peak error of a third-order loop in response to any of the three input variations discussed here would be about the same as that of a second-order loop with the same bandwidth and similar positions of the dominant poles. The differences arise in the steady-state errors and in the detailed transient behavior.

Sinusoidal Modulation

Next, let us investigate loop behavior in the presence of an angle-modulated input signal. For sinusoidal phase modulation

$$\theta_i(t) = \Delta\theta \sin \omega_m t \qquad\qquad (4.9)$$

and for sinusoidal frequency modulation

$$\theta_i(t) = \frac{\Delta\omega}{\omega_m} \cos \omega_m t \qquad\qquad (4.10)$$

Table 4.1 Transient Phase Error of Second-Order Loop, $\theta_e(t)$ (in rad) (high loop gain; $K_o K_d >> \omega_n$)

	Phase Step ($\Delta\theta$ rad)	Frequency Step ($\Delta\omega$ rad/sec)	Frequency Ramp ($\Delta\dot\omega$ rad/sec^2)
$\zeta < 1$	$\Delta\theta\left(\cos\sqrt{1-\zeta^2}\,\omega_n t - \dfrac{\zeta}{\sqrt{1-\zeta^2}}\sin\sqrt{1-\zeta^2}\,\omega_n t\right)e^{-\zeta\omega_n t}$	$\dfrac{\Delta\omega}{\omega_n}\left(\dfrac{1}{\sqrt{1-\zeta^2}}\sin\sqrt{1-\zeta^2}\,\omega_n t\right)e^{-\zeta\omega_n t}$	$\dfrac{\Delta\dot\omega}{\omega_n^2} - \dfrac{\Delta\dot\omega}{\omega_n^2}\left(\cos\sqrt{1-\zeta^2}\,\omega_n t \;+\; \dfrac{\zeta}{\sqrt{1-\zeta^2}}\sin\sqrt{1-\zeta^2}\,\omega_n t\right)e^{-\zeta\omega_n t}$
$\zeta = 1$	$\Delta\theta(1-\omega_n t)e^{-\omega_n t}$	$\dfrac{\Delta\omega}{\omega_n}(\omega_n t)e^{-\omega_n t}$	$\dfrac{\Delta\dot\omega}{\omega_n^2} - \dfrac{\Delta\dot\omega}{\omega_n^2}(1+\omega_n t)e^{-\omega_n t}$
$\zeta > 1$	$\Delta\theta\left(\cosh\sqrt{\zeta^2-1}\,\omega_n t - \dfrac{\zeta}{\sqrt{\zeta^2-1}}\sinh\sqrt{\zeta^2-1}\,\omega_n t\right)e^{-\zeta\omega_n t}$	$\dfrac{\Delta\omega}{\omega_n}\left(\dfrac{1}{\sqrt{\zeta^2-1}}\sinh\sqrt{\zeta^2-1}\,\omega_n t\right)e^{-\zeta\omega_n t}$	$\dfrac{\Delta\dot\omega}{\omega_n^2} - \dfrac{\Delta\dot\omega}{\omega_n^2}\left(\cosh\sqrt{\zeta^2-1}\,\omega_n t \;+\; \dfrac{\zeta}{\sqrt{\zeta^2-1}}\sinh\sqrt{\zeta^2-1}\,\omega_n t\right)e^{-\zeta\omega_n t}$
	Steady-state error = 0	Steady-state error = $\dfrac{\Delta\omega}{K_v}$ (not included above)	Steady state error = $\dfrac{\Delta\dot\omega t}{K_v} + \dfrac{\Delta\dot\omega}{\omega_n^2}$ ($\Delta\dot\omega t/K_v$ not included above)

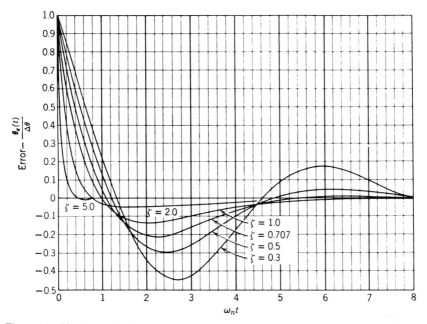

Figure 4.1 Phase error $\theta_e(t)$ due to a step in phase $\Delta\theta$. From Ref. 1 by permission of L. A. Hoffman.

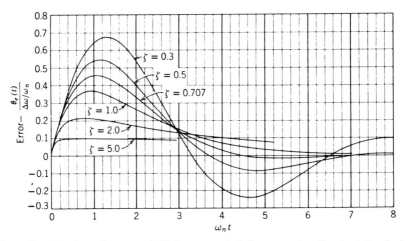

Figure 4.2 Transient phase error $\theta_e(t)$ due to a step in frequency $\Delta\omega$. (Steady-state velocity error, $\Delta\omega/K_v$, neglected.) From Ref. 1 by permission of L. A. Hoffman.

49

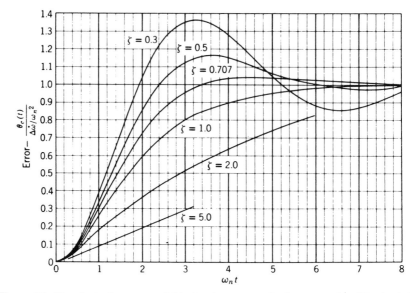

Figure 4.3 Transient phase error $\theta_e(t)$ due to a ramp in frequency $\Delta\dot{\omega}$. (Steady-state acceleration error, $\Delta\dot{\omega}/\omega_n^2$ included. Velocity error, $\Delta\dot{\omega}t/K_v$, neglected.) From Ref. 1 by permission of L. A. Hoffman.

where $\Delta\theta$ is peak phase deviation, $\Delta\omega$ is peak frequency deviation, and ω_m is modulating frequency.

Phase error is sinusoidal and may be calculated simply as the steady-state frequency response of (2.7). Examples are shown in Figures 4.4 to 4.6.

Error response to phase modulation is a highpass function of modulating frequency, as shown in Figure 4.4. At low frequencies, the response amplitude rises at $6n$ dB/octave for an nth-order loop. (This statement assumes perfect integrators in the loop filter. For real filters, the slope must break to 6 dB/octave at sufficiently low frequencies.) At high modulating frequencies, the loop is unable to follow the modulation so the full modulation phase appears as error at the phase detector. Accordingly, the high-frequency asymptote in Figure 4.4 is constant at 0 dB.

The curves of Figure 4.4 are sketches; the examples are different-order loops that have the same bandwidth in some sense. It is apparent that, for any frequency within the loop bandwidth, a higher-order loop tracks the modulation better than a lower-order loop does.

Error response to sinusoidal FM is shown in Figure 4.5 for three different kinds of loops. We note that the high-frequency asymptote is the same for all loops; the response differences lie at low frequencies within the loop bandwidth. The rolloff at high frequencies occurs solely because

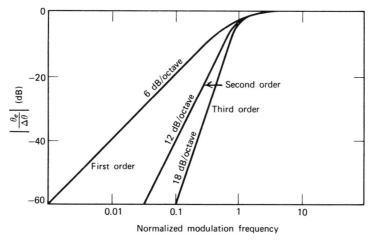

Figure 4.4 Phase error due to sinusoidal PM.

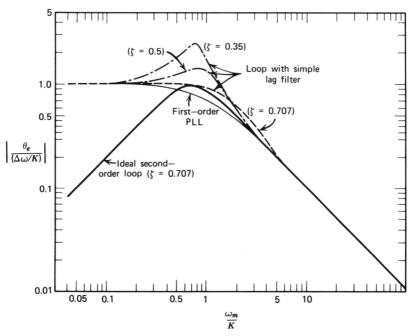

Figure 4.5 Phase error due to sinusoidal FM.

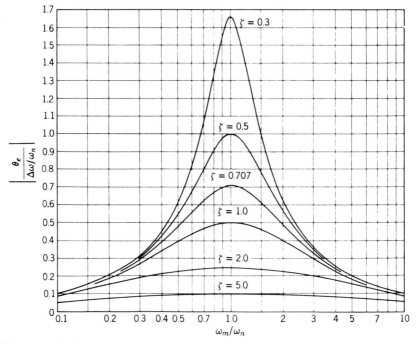

Figure 4.6 Steady-state peak phase error due to sinusoidal FM (high-gain, second-order loop). By permission of L. A. Hoffman.

input phase deviation $\Delta\theta = \Delta\omega/\omega_m$ is inversely proportional to modulating frequency.

The curves of Figure 4.5 have all been drawn for loops with equal loop gain K. We see that the first-order loop has a lowpass response in accordance with its one-pole transfer function, while the second-order loop is more effective at tracking out the lower frequencies.

A loop with a simple lag filter has the same asymptotes as those of a first-order loop with the same gain, but there is a response rise—error amplification—in the vicinity of the band edge, particularly for small damping factors. Performance of the lag loop, despite its second-order transfer function, is clearly much closer to a first-order loop than to that of true second order. If small phase error is needed, then a lag loop is likely to be a poor choice, even worse than a first-order loop.

Figure 4.6 shows phase error in response to FM, plotted against natural frequency and damping of a high-gain, second-order loop. Phase error is maximum at a modulating frequency equal to natural frequency ω_n, where peak amplitude is $\Delta\omega/2\zeta\omega_n = \Delta\omega/K$, irrespective of damping. Furthermore, the phase shift between phase error and the frequency-modulating wave passes through zero at $\omega = \omega_n$. These properties are sometimes used as the basis for experimental measurement of ω_n.

4.2 NONLINEAR TRACKING: LOCK LIMITS

All the preceding material on tracking and phase error is based on the assumption that the error is sufficiently small, thus allowing the loop to be considered linear in its operation. This assumption becomes progressively less useful as error increases until finally the loop drops out of lock, and the assumption becomes worthless. In this section the linear assumption is discarded, and the limiting conditions for which a loop holds lock are investigated.

Steady-State Limits

The first topic considered is the input frequency range over which the loop will hold lock. In (4.4a) the linear approximation of phase error due to a frequency offset is shown to be $\theta_v = \Delta\omega/K_v$. However, for a sinusoidal-characteristic phase detector the true expression should be

$$\sin\theta_v = \frac{\Delta\omega}{K_v} \tag{4.11}$$

The sine function cannot exceed unit magnitude; therefore, if $\Delta\omega > K_v$, there is no solution to this equation. Instead, the loop falls out of lock and the phase-detector voltage becomes a beat-note rather than a DC level. *Hold-in range* of a loop may therefore be defined as

$$\boxed{\Delta\omega_H = \pm K_v} \tag{4.12}$$

Equation (4.12) states that the hold-in range can be made arbitrarily large, simply by using very high loop gain. Of course, this cannot be entirely correct because some other component in the loop will then saturate before the phase detector; that is to say, to achieve any given frequency deviation of the VCO some definite control voltage is needed. However, the loop amplifier (if one is used) has some maximum voltage it can deliver and the VCO has some maximum voltage it can accept. If either of these limits is exceeded, the loop unlocks. It is not uncommon to find active loops with such high gain that the amplifier will saturate when static phase error is only a few degrees.

Dynamic error in a second-order loop was previously (4.7) approximated as $\theta_a = \Delta\dot{\omega}/\omega_n^2$. The correct expression for a phase detector with a sinusoidal characteristic should be

$$\sin\theta_a = \frac{\Delta\dot{\omega}}{\omega_n^2} \qquad (4.13)$$

from which it may be deduced that the maximum permissible rate of change of input frequency is

$$\boxed{\Delta\dot{\omega} = \omega_n^2} \qquad (4.14)$$

If the input rate should exceed this amount the loop will fall out of lock.

Many phase detectors have greater linear spans and larger maximum output than the sinusoidal characteristic of (4.11). Several examples are shown in Figure 4.7. All curves of Figure 4.7 are shown with the same slope at $\theta_e = 0$, which means that the different PDs all have the same gain factor K_d. Circuits that provide these and other extended characteristics are described in Chapter 6.

Increased PD output capability provides a larger tracking range—larger lock limit—than is obtainable from a sinusoidal PD. (Of course, the

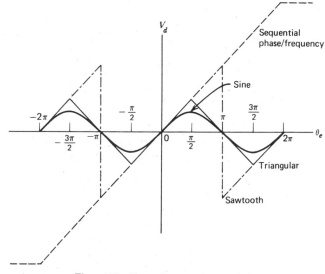

Figure 4.7 Phase-detector characteristics.

Table 4.2   Lock-Limit Extension

PD Type	Extension Factor
Sinusoidal	1
Triangular	$\dfrac{\pi}{2}$
Sawtooth	π
Sequential phase/frequency	2π

extended range of the PD is helpful only if the limit is set by the PD and not by some other nonlinearity, such as clipping in the operational amplifier.) Tracking extension of each of the types shown in Figure 4.7 is given in Table 4.2. Hold-in (4.12) and rate limits (4.14) are both extended by the same factor.

Transient Limits: Phase-Plane Analysis

Figure 4.2 demonstrates that transient error can be much larger than steady-state error, implying that a loop can be pulled out of lock on a transient basis by an input change that could be tracked easily in the steady state. This section examines a number of such transient conditions.

Most phase detectors are periodic and so cannot distinguish a phase step of $\Delta\theta + 2\pi n$ from one of $\Delta\theta$. Therefore, in the absence of other stress, an ordinary loop should never lose lock when subjected to a phase step, irrespective of magnitude or loop order.

A frequency step can break the lock. A first-order loop loses lock if, and only if, the frequency error exceeds the hold-in limit of (4.12), for a sinusoidal PD.

To study transient performance of a second-order loop, it is useful to introduce the *phase-plan portrait*. A description of phase planes, in general terms, is found in Truxal[2], and Viterbi[4] has specialized it to the PLL.

The dynamics of a second-order loop may be described by a pair of first-order, nonlinear differential equations in the independent variable of time and the dependent variables of phase error θ_e and frequency error $\dot{\theta}_e$. One can eliminate the time variable between the equations and arrive at a single, second-order, nonlinear differential equation that relates phase and frequency errors.

Solutions of the second-order equation are in terms of $\dot{\theta}_e$ versus θ_e; these can be plotted in the phase plane, which has θ_e and $\dot{\theta}_e$ as its coordinates. (Solutions cannot be obtained analytically; computer assistance is needed.) A plot of a single solution in the phase plane is known as a phase-plane *trajectory*. A family of trajectories is known as a phase-plane *portrait*. A trajectory shows the dynamic behavior of a loop as it settles (or fails to settle) towards equilibrium.

Figure 4.8 is a sketch of one particular phase-plane portrait for a sinusoidal phase detector and critical damping. Different portraits are obtained, depending on loop damping, phase-detector characteristic, loop stress, or signal modulation. The best source of portraits may be found in Viterbi's original report,[3] if accessible. His book[4] contains the same portraits, but at an inconveniently reduced scale. Blanchard's book[5] has a few of the portraits on a larger scale. Many of the specific results that follow in this section and in the next chapter were obtained by use of the portraits in Ref. 3. Phase-plane analysis is central to the understanding of nonlinear dynamics of second-order loops.

The phase-plane portrait of a PLL with periodic phase detector is itself periodic with period 2π in the variable θ_e, but it is aperiodic in $\dot{\theta}_e$. The pattern repeats indefinitely along the phase axis; two complete cycles are shown in Figure 4.8.

Trajectories proceed clockwise only, as shown by the flow arrows. Intersection of trajectories can occur only at *singular points*, which can be either stable or unstable. Equilibrium occurs at a stable singularity, which

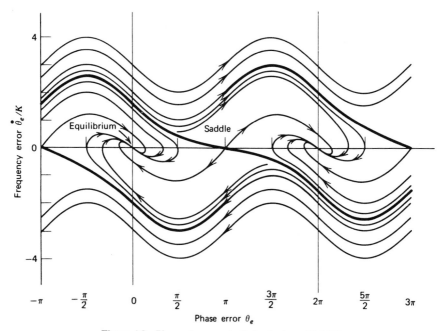

Figure 4.8 Phase-plane portrait, $\zeta = 1$, sinusoidal PD.

is called a stable node if the loop is overdamped or a stable focus if it is underdamped.[2] Equilibrium is the steady-state tracking condition reached after infinite time.

The unstable singularity is called a *saddle point*; the loop state cannot remain at a saddle point indefinitely because any slight disturbance sets it on an active trajectory.

A trajectory that terminates on a saddle point is called a *separatrix*. The separatices of Figure 4.8 are indicated by heavy curves. (The designation "separatrix" should apply only in the 2π interval in which the trajectory terminates on the saddle point and not all the way back into the infinite past.)

If a trajectory lies between the two separatrices, it will terminate at the equilibrium point of that particular 2π interval. If a trajectory lies outside the separatrices, the loop slips one or more complete cycles before arriving at equilibrium.

We are now ready to consider transients in a second-order loop. First, let us consider a loop with infinite DC gain. In principle, this kind of loop can never lose lock permanently. If a large frequency step is applied, the loop unlocks, skips cycles for a while, and then locks up once again. The phase error is a ringing oscillation for a number of cycles corresponding to the number of cycles skipped.

There is some frequency-step limit below which the loop does not skip cycles but remains in lock; we denote this limit as the *pull-out* frequency and give it the label $\Delta\omega_{PO}$. If we assume that the loop is at equilibrium at the instant the frequency step is applied, then the pullout limit is simply the intercept of the separatrix with the $\theta_e = 0$ axis. Using the portraits of Ref. 3, the pullout limit for a sinusoidal PD is found to have the values indicated in Figure 4.9. These data points fit the empirical relation

$$\Delta\omega_{PO} = 1.8\omega_n(\zeta + 1) \tag{4.15}$$

for ζ between 0.5 and 1.4.

The phase portrait can also be used to determine peak phase error for large steps of frequency. For $\Delta\omega = \Delta\omega_{PO}$, the peak phase error is 180°. However, the error increases rapidly as soon as it exceeds 90° and therefore the frequency step causing 90° peak error is only slightly less than $\Delta\omega_{PO}$. Figure 4.10 shows the situation for the special case of $\zeta = 0.707$.

Time history along any trajectory not too close to a singular point can be obtained from time mark curves called isochrones.[5] These are omitted from Figure 4.8 for clarity of explanation. It is also possible to obtain time intervals by graphical methods.[2]

A phase plane is applicable only to a second-order loop (or a degenerate phase plane to a first-order loop). A third-order loop has three state

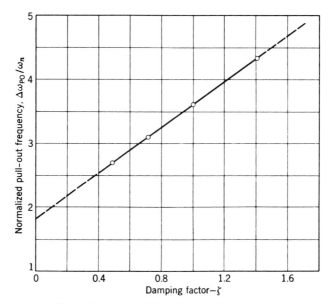

Figure 4.9 Pull-out frequency of high-gain, second-order loop. Sinusoidal PD.

variables—phase, frequency, and frequency rate—so it must have a three-dimensional phase *space* to represent it completely. Presentation of such a space is very difficult to achieve in two dimensions. As a result, we know much less about the transient response of third-order loops than we do about that of second-order loops. Viterbi,[3,4] Gupta,[8] and Tausworthe and Crow[9] have made attempts at solving the problem, but much more remains to be learned.

Modulation Limits

We must also be concerned with hold-in problems when the input signal is angle modulated. If the modulation index is excessive, a PLL is unable to remain locked to the signal.

It is useful to distinguish between *carrier tracking loops* in which the modulation spectrum is entirely outside the loop bandwidth and *modulation tracking loops* in which the modulation spectrum is inside the loop bandwidth. The first type is useful primarily for demodulation of small-index PM signals, while the second is needed to accommodate large-index FM or PM signals.

In the carrier-tracking loop the modulation must be restricted so that there actually is a carrier to track. If sinusoidal phase modulation of peak

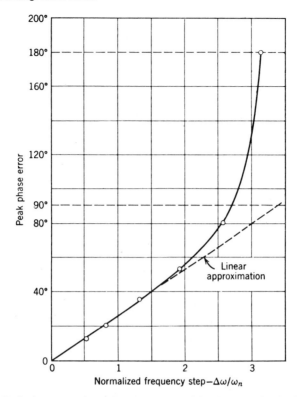

Figure 4.10 Peak phase error due to frequency step. High-gain, second-order loop; $\zeta = 0.707$. Sinusoidal PD.

deviation θ is applied, the carrier strength is proportional to the zero-order Bessel function $J_o(\theta)$. This function passes through its first zero for $\theta = 2.4$ rad (137°). Experiments have demonstrated that lock is lost very close to that first null.

As deviation is increased beyond 2.4 rad, lock is regained and is held until deviation reaches the next null of J_0 at $\theta = 5.5$ rad. In principle, a carrier-tracking loop loses lock on a sinusoidally modulated signal only in the immediate vicinity of the carrier nulls and holds lock for all other modulation indices.

Behavior of a modulation-tracking loop cannot be explained nearly so easily.[10] To introduce the problem, let us imagine a laboratory experiment in which a sinusoidally modulated signal (PM or FM) is applied to a PLL with sinusoidal phase detector. We require that the loop gain K be very much larger than the modulation frequency ω_m (that is, the modulation frequency is well inside the loop bandwidth; if otherwise, high-frequency effects intervene and the explanation becomes even more difficult).

The PD output voltage is observed on an oscilloscope and the modulation index is varied, starting at small deviation. At small deviation the observed wave shape is sinusoidal, as would be expected. Amplitude of the PD output increases with increasing deviation. If the deviation is made too large, the loop begins slipping cycles and severe distortion appears on the scope face (slip details are given later).

However, the PD output appears to remain nearly sinusoidal—nearly undistorted—from small index all the way up to the break-lock condition. This behavior is rather surprising, since we know that the PD operates well into its nonlinear region before break lock is reached. How can there be low-distortion operation in a nonlinear device? The answer, of course, is that negative feedback cancels out most of the distortion at the PD output, provided that loop gain is large at the modulation frequency. Reduction of distortion is a familiar property of feedback loops in general that is shared by the PLL in particular.

If the PD output is almost undistorted, then the peak phase error must increase as the inverse sine of the deviation, to a good approximation. In other words, we find that the distortion expected because of the PD nonlinearity appears in the phase error θ_e but not in the PD output $v_d = K_d \sin \theta_e$. This allows us to determine the wave shape of the phase error as a function of the peak PD output.

If the PD output is sinusoidal, then it must be of the form $v_d(t) = a K_d \sin \omega_m t$, where a is a factor between 0 and 1. Maximum possible output voltage from a sinusoidal phase detector is K_d V, so a is the ratio of peak output to maximum-possible output. Furthermore, $v_d(t) = K_d \sin \theta_e(t)$, from which we find (valid in the first quadrant)

$$\theta_e(t) = \arcsin(a \sin \omega_m t) \qquad (4.16)$$

Examples for several values of a are plotted in Figure 4.11. Considerable distortion of θ_e is evident for large values of a, but the curves for v_d are all sinusoidal.

Once it is recognized that v_d is substantially undistorted, it becomes a simple matter to establish the modulation limits. Knowing the input modulation, we can find v_d by modification of (2.7) to

$$V_d(s) = K_d[1 - H(s)]\theta_i(s) \qquad (4.17)$$

where $\theta_i(s)$ is the L-transform of the phase modulation on the signal. Taking the inverse L-transform of (4.17) yields $v_d(t)$, from which the peak value v_{dp} can be determined.

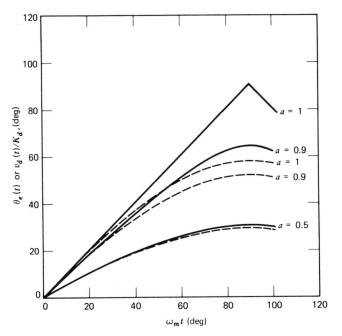

Figure 4.11 Phase error in sinusoidal PD. Input signal is sinusoidally angle modulated with frequency ω_m. (—) phase error $\theta_e(t)$. (---) PD output, $v_d(t)/K_d$. $v_d(t) = aK_d \sin \omega_m t = K_d \sin \theta_e(t)$.

The loop remains in lock if $v_{dp} < K_d$ and slips cycles if the calculated value of $v_{dp} > K_d$. (This criterion is applicable to a sinusoidal PD and must be modified for other PD characteristics.)

Maximum-possible output of the sinusoidal PD is K_d volts. Feedback maintains v_d almost undistorted right up to the maximum and then the loop fails abruptly as its capability is exceeded.

If modulation is sinusoidal with modulating frequency ω_m and peak frequency deviation $\Delta\omega$, then the deviation limit is found to be[10]

$$\Delta\omega = K \qquad \text{(first-order loop; } \omega_m \ll K) \qquad (4.18a)$$

$$\Delta\omega = \frac{\omega_n^2}{\omega_m} \qquad \text{(second-order loop; } \omega_m \ll \omega_n) \qquad (4.18b)$$

Modulation limits have also been worked out for frequency modulation by a gaussian message[10] with a baseband spectrum flat from DC to a cutoff frequency B_m Hz and rms deviation σ_f Hz. Gaussian signals have unbounded peaks, so there is some small probability that the loop will occasionally slip a cycle no matter how small the rms deviation. As an engineering tool, we invoke the concept of *crest factor* and give it the symbol γ. We choose γ such that the instantaneous deviation is less than $\gamma\sigma_f$ almost all the time. A value of $\gamma = 3.5$ has been found as a good empirical fit to laboratory observations of modulation unlock.

Using these concepts, the lock limits for gaussian modulation are found to be[10]

$$\sigma_f = \frac{K}{2\pi\gamma} \quad \left(\text{first-order loop; } B_m \ll \frac{K}{2\pi}\right) \tag{4.19a}$$

$$\sigma_f = \frac{\sqrt{3}\,\omega_n^2}{4\pi^2\gamma B_m} \quad \left(\text{second-order loop; } B_m \ll \frac{\omega_n}{2\pi}\right) \tag{4.19b}$$

Detailed behavior of the loop at the unlock threshold is rather curious.[11] If modulation is sinusoidal, the PD output remains virtually sinusoidal for any deviation up to the lock limit. An infinitesimal increase beyond the limit causes a drastic change in the PD output.

For a first-order loop (Figure 4.12) large spikes suddenly appear. Each spike represents the slip of one cycle. Slip spikes occur only while the instantaneous deviation is beyond the lock limit and the loop relocks as soon as the deviation returns within the loop bounds. A single spike appears for each modulation peak if the overmodulation is slight, while additional spikes appear in bursts as the overmodulation is increased.

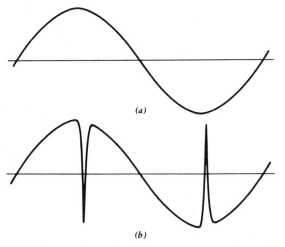

(a)

(b)

Figure 4.12 Modulation waveforms from PD: (a) in lock; (b) first-order PLL, unlocked.

Behavior of a second-order loop is quite different. Phase-detector output is sinusoidal right up to the lock limit, but an infinitesimal increase of deviation causes the loop to go completely out of lock and only a beat-note appears at the PD output. It is impossible to adjust the deviation to obtain the single spikes of Figure 4.12. (These statements apply only if the modulation frequency is much less than the natural frequency of the loop.) Why should a second-order loop behave so differently from a first-order loop?

The answer is not entirely clear, but some insight is gained when we realize that the peak phase error does not occur at the peak of the frequency-modulation cycle (as in the first-order loop). Instead, peak phase error coincides with maximum rate of change of frequency, which corresponds to zero instantaneous deviation for sinusoidal modulation. In fact, the sinusoidal unlock criterion (4.18b) can be shown to be the same as $\Delta\dot\omega = \omega_n^2$, which is the sweep-rate limit for a second-order loop (4.14).

Why does the loop not relock as soon as the modulation cycle has passed through the region of excessive frequency rate? (After all, the first-order loop relocks as soon as the region of excessive frequency deviation has been passed). That question has not yet been answered satisfactorily. However, the signal and the VCO are in frequency agreement only at the instant of maximum frequency rate and the loop cannot lock there if the rate is excessive. At other points in the cycle, the rate is lockable, but the frequency difference between signal and VCO prevents locking. Chapter 5 provides a better appreciation of this matter, but there is room for an improved explanation.

As a final comment, note that the feedback of the PLL makes it attractive as an FM demodulator. The feedback causes the baseband-recovered signal to remain largely undistorted right up to unlock, even though the phase detector itself is a nonlinear device and the phase error is distorted. Further discussion of FM demodulators is found in Chapter 9.

REFERENCES

1. L. A. Hoffman, *Receiver Design and the Phase-Lock Loop*, Aerospace Corporation, El Segundo, CA, May 1963.
2. J. G. Truxal, *Automatic Feedback Control System Synthesis*, McGraw-Hill, New York, 1955, Chap. 11.
3. A. J. Viterbi, *Acquisition and Tracking Behavior of Phase-Locked Loops*, External Publication No. 673, Jet Propulsion Laboratory, Pasadena, CA, July 1959.
4. A. J. Viterbi, *Principles of Coherent Communication*, McGraw-Hill, New York, 1966, Chap. 3.
5. A. Blanchard, *Phase-Locked Loops*, Wiley, New York, 1976, Chap. 10.

6. P. H. Lewis and W. E. Weingarten, "A Comparison of Second, Third, and Fourth Order Phase-Locked Loops," *IEEE Trans.*, *AES-3*, pp. 720–727, July 1967.

7. S. L. Goldman, "Jerk Response of a Third-Order Phase-Lock Loop," *IEEE Trans.*, *AES-12*, pp. 293–295, March 1976.

8. S. C. Gupta, "Transient Analysis of a Phase-Locked Loop Optimized for a Frequency Ramp Input," *IEEE Trans.*, *SET-10*, pp. 79–83, June 1964.

9. R. C. Tausworthe and R. B. Crow, *Practical Design of Third Order Phase-Locked Loops*, JPL Report 900–450, Jet Propulsion Laboratory, Pasadena, CA, April 1971.

10. F. M. Gardner and J. F. Heck, "Angle Modulation Limits of a Noise-Free Phase Lock Loop," *IEEE Trans.*, *COM-26*, pp. 1129–1136, August 1978.

11. F. M. Gardner and J. F. Heck, "Phaselock Loop Cycle Slipping Caused by Excessive Angle Modulation," *IEEE Trans.*, *COM-26*, pp. 1307–1309, August 1978.

Chapter Five

Acquisition

In all the preceding chapters, it is assumed tacitly that the loop is already in lock. But a loop starts out in an unlocked condition and must be brought into lock, either by its own natural actions or with the help of auxiliary circuits. The process of bringing a loop into lock is called *acquisition* and is the subject of this chapter.

5.1 CHARACTERIZATION

If the loop acquires lock by itself, we call the process *self-acquisition* and if it is assisted by auxiliary circuits, the process is called *aided acquisition*. Self-acquisition is often a slow and unreliable process. Although a PLL is an excellent tracking device, it tends to be rather clumsy in acquisition. Therefore, acquisition-aid circuits are commonly used and it is not unusual to find them constituting half of the total circuitry in representative PLLs.

An nth-order loop contains n integrators, which can be ideal (such as the VCO), near-ideal (as in an active filter), or imperfect approximations (as in a passive filter). With each integrator there is associated a state variable of the loop: phase, frequency, frequency rate, and so on. To bring the loop into lock, it is necessary to set each of the state variables—each of the integrators—to be in close agreement with the corresponding parameters of the input signal. Therefore, we should speak of phase acquisition, frequency acquisition, and so forth, up to n forms of acquisition for an nth-order loop. (In most instances we restrict our attention to just two or three state variables; some of the storage elements in the loop contribute only high-frequency poles and can be neglected.) Frequency acquisition has received the most attention, but the other state variables are also important, sometimes critically so.

Acquisition is inherently a nonlinear phenomenon; we must use nonlinear analysis throughout and cannot ease the way with linear approximations.

5.2 PHASE ACQUISITION

Under ordinary conditions, phase is self-acquired. Study of phase acquisition leads to better understanding of the overall acquisition problem and provides guidance if aided phase acquisition is needed.

First-Order Loop

It is instructive to begin with analysis of a first-order loop. To show performance, we derive the nonlinear differential equation of the loop and analyze its meaning.

We let ω_i be the input frequency (assumed constant) and let ω_o be equal to the free-running frequency of the VCO so that the instantaneous frequency of the VCO is $\omega_o + K_o v_d$. Voltage $v_d = K_d \sin \theta_e$ is the error voltage out of the phase detector.

Input phase is $\omega_i t$ and output phase is

$$\theta_o = \omega_o t + \int_0^t K_o v_d \, dt + \theta_o(0)$$

$$= \omega_o t + \int_0^t K_o K_d \sin \theta_e \, dt + \theta_o(0) \tag{5.1}$$

Phase error is

$$\theta_e = \theta_i - \theta_o = \omega_i t - \omega_o t - \int_0^t K \sin \theta_e \, dt - \theta_o(0) \tag{5.2}$$

We let $\omega_i - \omega_o = \Delta\omega$ and differentiate to obtain

$$\dot{\theta}_e = \Delta\omega - K \sin \theta_e \tag{5.3}$$

This is the nonlinear differential equation of the first-order phaselock loop. The loop is locked only if $\dot{\theta}_e$ is zero, by definition of lock. However, we must determine whether the converse is true: that if $\dot{\theta}_e = 0$ the loop is necessarily locked.

The hold-in limit is obtained directly from (5.3); if $\dot{\theta}_e = 0$, then $\sin \theta_e = \Delta\omega / K$. Because $\sin \theta_e$ cannot exceed unity, the loop can lock only if $\Delta\omega / K < 1$.

It is useful to divide (5.3) by K and then plot $\dot{\theta}_e / K$ versus θ_e, as in Figure 5.1.* From the figure it may be seen that, if $|\Delta\omega / K| < 1$, there are two

---

*This analysis follows a similar one by Viterbi.[1,2] Figure 5.1 is a degenerate phase-plane portrait.

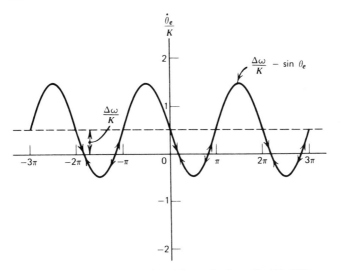

Figure 5.1   Phase-plane plot of first-order loop ($\Delta\omega/K = 0.5$).

points (nulls) in each interval of 2π for which $\dot{\theta}_e$ goes to zero. At a null the frequency difference between input and VCO is zero.

Adajacent nulls are of opposite slope. To analyze the behavior of the loop, let us suppose that the operating point is slightly displaced from one of the nulls. For one of negative slope the sign of $\dot{\theta}_e$ drives θ_e toward the null. (As an example, if phase displacement is slightly negative from a negative-slope null, the sign of $\dot{\theta}_e$ is positive and θ_e must necessarily increase—in the direction toward the null.) Conversely, a displacement from one of the positive-slope nulls will drive the state of the loop away from the null. Thus the negative-slope nulls are stable and the positive-slope nulls are unstable. Arrows show the direction of phase change.

Prior to lock $\dot{\theta}_e$ is nonzero, which means that θ_e must change (increase or decrease) monotonically. For this reason θ_e must eventually take on the value of one of the stable nulls (provided, of course, that $\Delta\omega < K$). When θ_e reaches a stable null, the loop is locked and θ_e remains fixed at the static error.

Because every cycle has a stable null, θ_e cannot change by more than one cycle before locking. Thus there is no cycle skipping in the lock-up process. The time required to lock up depends on the intial values of phase and frequency, but, as a rough rule of thumb, it will be on the order $3/K$ sec.

Exact settling time can be found[3] by integration of the differential equation 5.3. (Exact closed-form integration is possible for the first-order

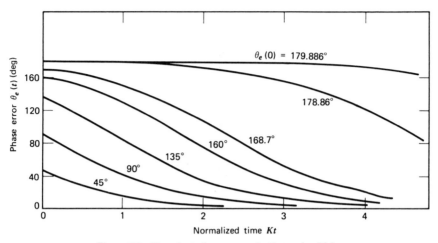

Figure 5.2 Transient phase errors in first-order PLL.

loop but not for second or higher orders.) Some example phase transients are shown in Figure 5.2 for $\Delta\omega=0$ and several values of $\theta_e(0)$. If θ_e is small, the loop operation is almost linear and the phase waveforms are nearly exponentials with time constant $1/K$. If θ_e is large, the waveforms diverge substantially from a simple exponential and settling times increase from those attained by an exponential of the same initial phase error.

If the initial phase error is very close to the unstable null, the phase can dwell near the null for an extended time, as illustrated by the two upper curves of Figure 5.2. This dwell phenomenon has been dubbed the *hangup effect*[4] and can be extremely troublesome in applications where rapid acquisition is needed with high reliability.

(Hangup is illustrated for a noisefree, first-order loop with sinusoidal phase detector and zero frequency error. Despite intuitive notions to the contrary, changing any or all of these conditions does not eliminate hangup. Specifically, hangup is aggravated by noise or other disturbances; second- or higher-order loops are equally subject to hangup; using an extended phase-detector characteristic (e.g., sawtooth) can alleviate hangup, but not eliminate it; and offsetting the frequency merely shifts the location of the unstable null, as shown in Figure 5.1. The full causes of hangup and some remedies are presented in Ref. 4.)

Lock-In

If signal frequency is close enough to VCO frequency, a PLL locks up with just a phase transient; there is no cycle slipping prior to lock. The frequency range over which the loop acquires phase without slips is called the *lock-in range* of the PLL.

In a first-order loop, the lock-in range is equal to the hold-in range; the loop self-acquires any signal that it can hold.

The same is not true of second- or higher-order loops; the lock-in range is invariably less than the hold-in range. Moreover, there is a frequency interval, smaller than the hold-in interval and larger than the lock-in interval, over which the loop will acquire lock after slipping cycles for a while. This intermediate interval is called the *pull-in range* and is discussed in the next section.

Lock-in is self-acquisition of phase by a PLL and is the subject of this section.

To assure stable tracking, it is common (though not universal) practice to build loop filters with equal numbers of poles and zeros. The lag–lead filter for the familiar second-order loop (see Chapters 2 and 4) has one pole and one zero. An inherent property of any such filter is that its frequency response is asymptotically flat, with zero phase, at high frequencies. An example for the second-order loop is sketched in Figure 5.3.

We denote the high-frequency asymptotic response by $F(\infty)$. In the lag–lead filter $F(\infty) = \tau_2/\tau_1$ irrespective of whether the filter is active or passive and, if active, irrespective of the DC gain of the amplifier.

At high frequencies the loop is indistinguishable from a first-order loop with gain $K = K_o K_d F(\infty)$. As a fair approximation, we can say that the higher-order loop has the same lock-in range as the equivalent-gain, first-order loop.

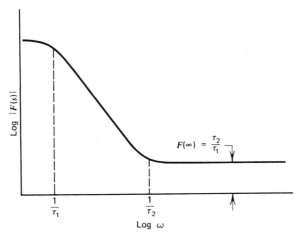

Figure 5.3 Frequency response of loop filter.

Loop gain is introduced in Section 2.3. For any loop whose filter has equal numbers of poles and zeros the gain is defined as $K = K_o K_d F(\infty)$— which is exactly the gain of the equivalent first-order loop. In fact, the definition of loop gain was deliberately chosen to make this relation be true.

The lock-in limit of a first-order loop is equal to the loop gain. We argue here that a higher-order loop has nearly the same lock limit. Denoting the lock limit as $\Delta\omega_L$,

$$\boxed{\Delta\omega_L \simeq \pm K} \tag{5.4}$$

is a useful engineering approximation for lock-in range.

Equation 5.4 is obtained under assumption of a sinusoidal phase-detector characteristic. Presumably, an extended PD characteristic (Figure 4.2) would also extend the lock limit.

The argument leading to (5.4) is a simplification of the real behavior of a PLL. In a higher-order loop it is not possible to determine whether the loop will or will not slip cycles, before locking, on the basis of initial frequency error alone; all initial state variables must be examined. In a second-order loop, the variables are frequency and phase; they are studied with the aid of a phase-plane portrait.

When inspecting the phase plane (e.g., Figure 4.8), we see immediately that the whole concept of lock-in is oversimplified. A second-order loop locks without slips if the initial state falls between the separatrices. Since a separatrix is a sinuous boundary, there is no natural way to define exactly any unique lock-in frequency.

One might arbitrarily define the average ordinate of the positive separatrix as the lock-in frequency. Or, the definition might be the separatrix ordinate at $\theta_e = 0$ or at $\theta_e = -180°$. Examination of Figure 4.8, or the more numerous set of portraits in Ref. 1, suggests that (5.4) is a conservative estimate of lock-in range.

Despite its vague reality, lock-in range is a useful concept for engineering calculations and in analyses presented in later paragraphs.

Aided Phase Acquisition

Unless hangup is a problem, phase usually is self-acquired if the phase detector has any of the usual characteristics (e.g., Figure 4.7). However, there are some signal types for which the phase-detector characteristic has only a small active region; over most of the phase-error interval, the PD output is zero. An example is shown in Figure 5.4.

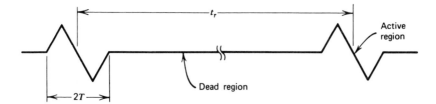

Figure 5.4 Phase-detector characteristic for gated-pulse or PRN signal. $\delta = 1$ (PRN signal); $\delta = T/t_r$ (pulsed signal); $T =$ pulse or chip width; $t_r =$ pulse or code repetition interval.

A pseudo-random noise (PRN) signal[5] is one kind that yields such a PD characteristic, and a gated pulse train is another. (The phase detector for the latter might be a radar range gate.)

A loop of this sort can acquire phase only if the initial phase error falls into the active region of the PD. If the initial error lies in the dead portion of the PD characteristic, there is no error information of any kind available to the loop, so acquisition can occur only by accident of phase drift. Likelihood of acquisition would be very poor if the PRN code were long or if the pulse duty cycle were short.

To assure acquisition, the equipment performs a phase search over all phases. When the active region of the PD is encountered, the loop locks up and the search is discontinued. Application of a phase search constitutes aided acquisition of phase.

A continuous phase sweep is the same as a frequency offset in the VCO and is usually an easy way to implement a phase search. If the phase rate (frequency offset) is too large, the search will sweep right through the active region without stopping and go on into the next dead region. There is a rate limit that must not be exceeded if acquisition is to be successful.

Acquisition with a second-order loop is analyzed by means of a phase-plane portrait. Gilchriest[6] has investigated the PRN signal and Gardner (unpublished) has examined the gated pulse train. They have found, for a PD characteristic of the kind shown in Figure 5.4 and for damping factors of 0.75 or greater, that the maximum phase rate is given by

$$\Delta f \simeq B_L \delta \qquad \text{cycles/sec} \qquad (5.5)$$

where B_L is the noise bandwidth defined in Chapter 3, and δ is the pulse duty ratio. For the PRN signal, $\delta = 1$. As might be expected, changing the shape of the PD characteristic has substantial influence on the allowable phase sweep rate. Moreover, smaller damping reduces the allowable rate.

5.3 FREQUENCY ACQUISITION

Acquisition of frequency ordinarily is more difficult, is slower, and requires more design attention than does phase acquisition. In consequence, the literature has concentrated largely on frequency acquisition, to the point that "acquisition" is almost synonymous with "frequency acquisition." Furthermore, the study of frequency acquisition has been devoted mainly to the second-order loop, partly because of its technological importance, but also because of the greater difficulties of analyzing higher-order loops. Discussion in this section concentrates mostly on second-order loops.

Self-acquisition of frequency is known as *frequency pull-in*, or simply, *pull-in*. Pull-in tends to be slow and often unreliable, so a number of aided frequency-acquisition techniques have been devised, including *frequency sweeping, frequency discriminators*, and *bandwidth-widening* methods.

Pull-In

Pull-in, particularly in a loop with very narrow bandwidth, is fascinating to watch. When the signal is first applied, the loop is not locked and only a beat-note appears at the output of the PD. Frequency of the beat-note slowly decreases—the VCO frequency slowly approaches that of the signal —until the lock limit is reached, whereupon the loop snaps into lock without any further cycle slipping.

Pull-in behavior may be understood by recognizing that the beat-note is reduced in amplitude by the loop filter but is not suppressed completely. An attenuated beat-note, with peak amplitude $K_d F(\infty)$ is applied to the VCO control terminal, causing the VCO to be frequency modulated at the beat frequency. (Throughout this analysis we assume that the PD has a sinusoidal characteristic and the loop filter has constant response at high frequencies, as in Figure 5.3.) Therefore, the PD output is the low-frequency multiplier product of a sine wave and a frequency-modulated wave. Since the modulating frequency is equal to the beat frequency, the beat-note waveform could hardly be sinusoidal.

Richman[3] has derived the waveform of the beat-note for a first-order loop by integrating the differential equation (5.3) of the loop. The explicit equation describing the waveform is cumbersome and does not provide much insight into the problem. However, a plot of the wave-form, as in Figure 5.5, is very revealing, and the nonsinusoidal character of the beat-note is evident. Moreover, and vitally important, the positive and negative excursions are obviously unequal in area; therefore, the phase-detector output must contain a DC component even before lock is obtained. It is the presence of this component that allows pull-in to occur.

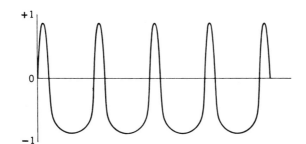

Figure 5.5 Typical beat-note wave shape, first-order loop, $\Delta\omega/K = 1.10$.

Once the existence of a DC component is recognized, an alternative explanation of its presence aids understanding; that is, the beat-note, of fundamental frequency $\Delta\omega$, frequency modulates the VCO whose center frequency is ω_o. This modulation generates FM sidebands in the VCO output at frequencies $\omega_k = \omega_o + k\Delta\omega$, where k takes on all integer values. In the phase detector this modulated output is multiplied by the sinusoidal input with frequency ω_i.

The difference signal out of the phase detector consists of individual signals at all the frequencies $\omega_i - \omega_k = \omega_i - \omega_o - k\Delta\omega$. Recall, however, that $\Delta\omega = \omega_i - \omega_o$ and, therefore, $\omega_i - \omega_k = \omega_o - k(\omega_i - \omega_o) = (1-k)(\omega_i - \omega_o)$. The individual signal corresponding to $k=1$ has a frequency of zero, that is, $k=1$ corresponds to a DC component. Relevant spectra are shown in Figure 5.6.

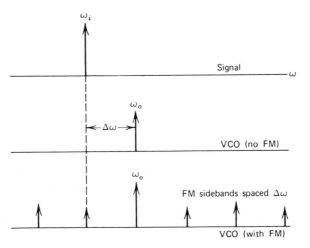

Figure 5.6 Pull-in spectra.

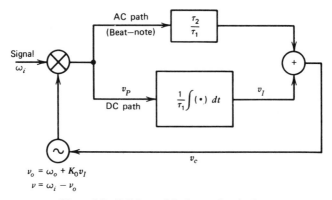

Figure 5.7 Pull-in model of second-order loop.

We give this DC component the name *pull-in voltage* and denote it by the symbol v_p.

In a first-order loop the effect is not of much value; if the initial difference frequency exceeds the lock-in frequency the magnitude of the DC component is insufficient to pull into lock. However, the average difference frequency is reduced; even the first-order loop tends to pull toward lock, despite the fact that it will not reach lock.

The second-order loop includes an integrator in its loop filter. This integrator builds up an increasing output in response to a DC input; the accumulated output (delivered to the VCO) can greatly exceed the magnitude of the filtered beat-note that modulates the VCO. As the integrator output builds up, the VCO frequency is adjusted toward the direction of lock. If the initial difference frequency is not too great, the loop will eventually lock up.

Approximate formulas for pull-in time and for pull-in limits may be obtained by following a method originated by Richman.[3] We consider the loop to be representable as in Figure 5.7. There is a high-frequency path from PD to VCO with flat gain of τ_2/τ_1 and a low-frequency path that contains an integrator. For the present, we regard the integrator as perfect.

Output of the phase detector consists of an AC beat-note and the DC pull-in voltage v_p. For analysis purposes, we pretend that the AC portion passes only through the high-frequency path and is completely suppressed in the integrator path. (There is little pretense involved for high-enough beat frequencies.) Likewise, we assume that the DC pull-in voltage is passed mainly by the integrator and only a negligible portion goes through the high-frequency path. This is an accurate approximation for time intervals appreciably larger than the time constant τ_2.

Input frequency is ω_i, free-running frequency of the VCO is ω_o, and initial frequency difference is $\Delta\omega = \omega_i - \omega_o$. Average frequency (average over a beat cycle) of the VCO during pull-in is denoted ν_o and is given by $\nu_o = \omega_o + K_o v_I$, where v_I is the output of the integrator. Any change in v_I or ν_o is negligible over the time of a single beat cycle. Average frequency error over a short time is $\nu = \omega_i - \nu_o$. If the loop is to pull in, rather than lock in, we require $|\Delta\omega| > K$.

Pull-in voltage is a function of ν. Richman has integrated the differential equation of a first-order loop and found its pull-in voltage to be

$$v_p = K_d \left[\frac{\nu}{K} - \sqrt{\left(\frac{\nu}{K}\right)^2 - 1} \; \right] \qquad (5.6)$$

for $|\nu| > K$. (In a first-order loop, $\nu = \Delta\omega$ since the integrator is omitted.) We use the same formula for pull-in voltage of a second-order loop, a reasonable expedient under the assumptions that have been imposed.

Combining the various equations about the low-frequency loop gives

$$\nu = \Delta\omega - \frac{K_o}{\tau_1} \int_0^t v_p \, dt \qquad (5.7)$$

which is differentiated to give the equation

$$\frac{d\nu}{dt} = - \frac{K_o v_p}{\tau_1} \qquad (5.8)$$

We substitute (5.6) for v_p and separate differentials to obtain

$$dt = - \frac{\tau_2 \, d\nu}{K\left[\dfrac{\nu}{K} - \sqrt{(\nu/K)^2 - 1} \; \right]} \qquad (5.9)$$

recognizing that $K_o K_d / \tau_1 = K / \tau_2$.

Pull-in time T_p is defined as the time required for the average frequency error to change from the initial condition $\nu = \Delta\omega$ to the lock limit $\nu = K$. We find T_p by integrating (5.9) between the limits of $\Delta\omega$ and K.

If we assume that $\Delta\omega \gg K$, then the result is

$$T_p \simeq \frac{(\Delta\omega)^2 \tau_2}{K^2} = \frac{(\Delta\omega)^2}{2\zeta\omega_n^3} \qquad (5.10)$$

Because of the approximations, this formula should not be applied if $\Delta\omega$ is either very large (near $\Delta\omega_p$)* or very small (near $\Delta\omega_L$). It is best applied in the midrange and should be considered as the time required to pull in from the initial offset to a beat frequency equal to $\Delta\omega_L$ (at which time the loop quickly locks in). For the special case of a high-gain loop with $\zeta = 0.707$, the pull-in time is

$$T_P = \frac{27(\Delta\omega)^2}{256 B_L^3} \approx \frac{4.2(\Delta f)^2}{B_L^3} \text{ sec} \tag{5.11}$$

A narrowband loop can take a very long time to pull in. For example, for a situation in which $\Delta f = 1 \text{ kHz}$ and $B_L = 10 \text{ Hz}$, pull-in time would be 1 hr and 10 min, which is intolerably long for almost any application.

If the loop filter contains a perfect integrator, pull-in will be accomplished no matter how large the initial frequency error. (This statement neglects clipping limits of the integrator; the loop clearly cannot pull in a signal that requires excessive control voltage to the VCO. Also, it is assumed that there are no unwanted DC offsets within the loop that would counteract the pull-in voltage and cause the VCO frequency to be pushed out instead.) In a real loop filter, the DC gain is some finite number $F(0)$, and if v_p is small enough—if the initial frequency error is large enough—the loop cannot pull in. The largest frequency for which the loop can still pull into lock is called the *pull-in limit* and is represented by $\Delta\omega_p$.

To derive the pull-in limit, we replace the perfect integrator in Figure 5.7 by an imperfect integrator with DC gain $F(0) - F(\infty)$. [The DC gain of the entire loop filter is $F(0)$, while the DC gain of the high-frequency path is $F(\infty)$. Therefore, the DC gain of the low-frequency path must be $F(0) - F(\infty)$.]

Let us assume that $|\Delta\omega| > \Delta\omega_p$ so that the loop cannot pull in. The phase detector still generates a pull-in voltage v_p, which is amplified by the factor $F(0) - F(\infty)$ and is applied to the VCO where it causes a steady-state frequency change of $K_0[F(0) - F(\infty)]v_p$. The steady-state frequency error is

$$\nu = \Delta\omega - K_0[F(0) - F(\infty)]v_p \tag{5.12}$$

Substituting (5.6) for v_p and remembering that $K_v = K_o K_d F(0)$ and $K = K_o K_d F(\infty)$ we obtain

$$\nu = \Delta\omega - (K_v - K)\left[\frac{\nu}{K} - \sqrt{\left(\frac{\nu}{K}\right)^2 - 1}\right] \tag{5.13}$$

---

*$\Delta\omega_p$ is the pull-in limit and is derived presently.

Equation 5.13 can be solved for the steady-state frequency error. A real solution is found if $\Delta\omega \geqslant K(2K_v/K - 1)^{1/2}$. A smaller value of $\Delta\omega$ leads to complex roots of (5.13), which means that no real final frequency error satisfies (5.13). We conclude that the loop pulls in for smaller values of $\Delta\omega$.

Because of the many approximations that have been made, the boundary is accurate only for a high-gain loop, in which $K_v \gg K$. Therefore, an approximate formula for pull-in limit is

$$\boxed{\Delta\omega_p \simeq \sqrt{2K_v K}} \qquad (5.14)$$

We see that, in principle, the pull-in range can be made as large as may be needed simply by using a large DC gain K_v. Moreover, the large pull-in can be achieved with as narrow a noise bandwidth as necessary; the two parameters are independent.

Formula 5.10 for pull-in time is valid only if the initial frequency error is substantially larger than the loop gain K and substantially smaller than the pull-in limit. Richman obtains improved formulas that describe the pull-in time for all conditions, including initial frequency error near either of the bounds. The results are much more cumbersome than (5.10).

A great many investigators have investigated pull-in. Viterbi[1,2] examined the problem through limit cycles in the phase plane and arrived at essentially the same results given here.

The foregoing results apply only to loops with sinusoidal phase detectors. Mengali[7] summarizes work by other authors on extended PD characteristics and arrives at general formulas for pull-in time and range that take the PD characteristic into account. As might be expected, an extended PD characteristic provides an extended pull-in range and faster pull-in time.

Meer[8] investigated extended PD characteristics and, also, higher-order loops. He derived the pull-in voltages associated with triangular and sawtooth PDs and observed that these are larger than for sinusoidal PDs.

In a third-order loop, there are two integrators in the low-frequency path; the double-integrated pull-in voltage has parabolic, rather than linear, growth. As a result, pull-in is faster in a third- or higher-order loop than it is for a second-order loop.

Let us assume that both integrators are ideal and that $\Delta\omega \gg K$. Following Meer's analysis, but using the notation of Section 2.3, we find pull-in time for a third-order loop to be

$$T_p \simeq \sqrt{\frac{\pi}{a_3}} \, \frac{\Delta\omega}{K} \qquad (5.15)$$

(If both zeros of the loop filter are coincident at $s = -1/\tau_2$, then $1/\sqrt{a_3} = \tau_2$.) Pull-in time for the third-order loop varies as the first power of initial frequency error, rather than $\Delta\omega^2$ as in a second-order loop (5.10).

Unfortunately, frequency pull-in to zero beat does not assure rapid phaselocking in a third-order loop. Equation 5.15 indicates the time needed to accumulate the correct tracking charge on the frequency integrator in the loop filter, but the charge on the frequency-rate integrator will be wrong at that time. It is entirely likely that the stored charge on the first integrator will force the second integrator to continue to charge, rather than stop at the proper frequency. If that should happen, the VCO frequency overshoots the correct equilibrium, pull-in voltage reverses polarity, and the pull-in action heads for equilibrium from the opposite direction.

In other words, the approach to lock can be oscillatory and (5.15) only tells the time to the first passage through zero frequency error, not to phaselocking. Lock is not possible until the charge on the first integrator dwindles to the correct value needed for equilibrium tracking.

On the other hand, in the vicinity of zero frequency error, the high-frequency path through the loop filter has a strong locking action. If that locking force can overcome the frequency-slewing force from the first integrator, then the loop will lock at first passage and will not oscillate about frequency equilibrium. The work of Tausworthe and Crow[9] suggests that lock occurs on first passage if the closed-loop poles are overdamped and oscillatory acquisition occurs if the poles are underdamped. Further investigation is needed to gain a better understanding.

The pull-in limit analysis leading to (5.14) depends only on the DC and AC loop gains and not on the loop order. Therefore, pull-in range should be independent of order, a deduction that is corroborated in Meer's paper.

The analyses and references presented above deal only with high-gain loops having loop filters with equal numbers of poles and zeros. The analyses fail badly if these conditions are violated. Chapter 8 shows some unfortunate consequences of additional poles within the loop.

Greenstein[10] examines pull-in range (but not time) for low-gain, second-order loops and also the loop with a simple lag filter. He avoids analytical approximations by processing the exact loop equations on a computer. His results agree with those of the other investigators for high-gain loops—the condition of most practical interest—and are much more accurate for low-gain loops.

Pull-in can be a painfully slow process in a narrowband loop, so there have been several attempts to speed it up. Hiroshige[11] has devised nonlinear switching circuits that allow more frequency-error reduction per beat cycle than is achieved in an ordinary loop. Pull-in time is reduced substantially, but there has been little practical application of the technique, perhaps because more potent methods, which use the same amount of extra circuitry, are available.

Runge[12] has found that a combination of phaselocking and injection locking improves the pull-in behavior. (An oscillator is injection locked by adding the incoming signal directly into the oscillator's tuned circuit. Injection locking is described by a nonlinear differential equation identical to that for the first-order phaselocked loop.[13]) Signal injection can often be accomplished with minimal additional circuitry and may be attractive from that standpoint. In fact, if care is not taken in physical layout of the circuits, injection may be difficult to avoid.

From the many papers on the subject, a casual reader might get the impression that pull-in is the dominant applied method of frequency acquisition. Actually, one could argue that pull-in is more interesting than it is practical. Besides its slowness, pull-in can be defeated by unwanted, but unavoidable, DC offsets arising in the phase detector (Chapter 6) or it can be converted to push-out or false locking by excess poles or delays within the loop (Chapter 8). There is virtually no information available on pull-in behavior in the presence of significant noise.

In the author's experience, pull-in is practical only in a comparatively benign environment: where noise is small, bandwidth is large enough to permit rapid action, and the loop circuits are simple, so extra poles are avoided. In more challenging applications, pull-in is almost always found to be unsatisfactory or unusable and some form of aided acquisition is needed. Forms of aided frequency acquisition are discussed on the following pages.

Frequency Sweeping

Improved frequency acquisition can be attained by sweeping the frequency of the VCO, thereby searching for the signal frequency. If the search is applied correctly, the loop will lock up as the VCO frequency sweeps into coincidence with the signal. Lock-up inhibits further change of VCO frequency, so the sweep process is self-terminating.

From the earlier discussion on hold-in in the presence of a frequency ramp it should be evident that the sweep rate must not be excessive. We show earlier that the loop cannot hold lock if the sweep rate $\Delta\dot{\omega}$ exceeds ω_n^2. If a loop cannot *hold* lock on a signal, it certainly will be unable to *acquire* lock. Therefore, an absolute maximum limit on the allowable sweep rate is $\Delta\dot{\omega} < \omega_n^2$ (for a PD with sinusoidal characteristic).

Viterbi[1,2] has investigated acquisition problems by means of phase-plane trajectories. He discovered that acquisition is not certain, even if $\Delta\dot{\omega} < \omega_n^2$ and the loop is noisefree. If $\Delta\dot{\omega}$ becomes somewhat larger than $\omega_n^2/2$, there is a possibility that the VCO can sweep right through the input frequency

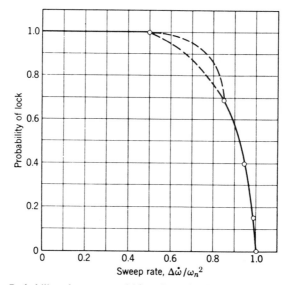

Figure 5.8 Probability of sweep acquisition. Second-order loop; $\zeta = 0.707$; no noise.

without locking. The chance of locking or nonlocking depends on the random initial conditions of frequency and phase. Viterbi's phase-plane trajectories were used to compute the probability of locking graphically, and it is plotted against sweep rate in Figure 5.8. These results apply directly only to the special case of a high-gain, second-order loop with $\zeta = 0.707$. However, qualitatively similar behavior should be expected for other damping factors.

Further qualitative information on sweep acquisition behavior is available from the simulation study[14] by Frazier and Page.* Their paper indicates that for fixed natural frequency and sweep rate the probability of lock improves as damping increases. See Figure 5.9, which seems to imply that the loop should be heavily damped, at least until it is locked.

Such a conclusion is premature; loop-noise bandwidth varies with damping even though natural frequency is fixed (refer to Figure 3.3). On the basis of fixed-noise bandwidth, the largest value of ω_n (and therefore the largest maximum sweep rate) occurs for $\zeta = 0.5$. Yet the probability of acquiring lock at sweep rates less than ω_n^2 improves as damping increases. There is some value of ζ that provides best acquisition performance; the exact value is not known, but it probably lies between 0.7 and 1.0.

---

*There is a numerical error by a factor of 1.4 that runs throughout their paper, making quantitative interpretations difficult.

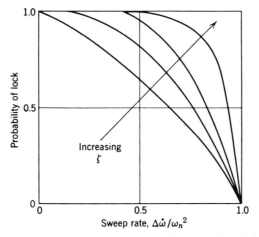

Figure 5.9 Probability of sweep acquisition showing effect of damping.

So far we have assumed that the loop is essentially noisefree. In actuality noise is always with us and must be taken into account. Simple intuition leads us to expect that noise will make it more difficult to acquire a signal; it would be useful if this difficulty could be expressed by a number.

Frazier and Page's experiments provide empirical data that suggest that sweep rate should be reduced by a factor of $[1-(SNR_L)^{-1/2}]$ if an acceptably high probability of acquisition is to be maintained in the presence of noise. This expression predicts that acquisition becomes impossible at 0 dB signal-to-noise ratio in the loop. Experience suggests this conclusion to be optimistic.

Combining the disparate fragments of information and the author's experience, a better preliminary design value for sweep rate might be

$$\Delta\dot\omega = \tfrac{1}{2}\omega_n^2\left[1-2(SNR_L)^{-1/2}\right] \tag{5.16}$$

in combination with $\zeta = 0.7$ to 1.0. This choice implies that sweep acquisition is impossible below 6-dB SNR_L, which is a somewhat conservative statement but not drastically wrong. Experimental adjustment from these values can provide refinement, if needed.

Because of the nonlinearity, sweep acquisition has defied satisfactory analysis in the presence of significant noise. For those who want more quantitative results than are given here, Blanchard[15] reports on an extensive series of laboratory measurements that relate sweep speed, signal-to-noise ratio, and probabilities of correct acquisition and false alarm.

The results given here and in the reference section apply to a loop with sinuoidal phase detector. A different PD characteristic can be expected to produce different sweep capabilities; the matter does not appear to have been investigated. (In Chapter 6, it is shown that a sinusoidal characteristic is usually the only kind possible if the input SNR is very small.)

Sweep can be applied to a second-order loop in a very simple and elegant manner. Some workers have built separate sawtooth generators that add a sweep voltage directly into the VCO, but this approach is unnecessarily complicated and arises from inadequate understanding of the state variables of the loop.

A far better approach is to insert a constant slewing current into the integrator of the loop filter. Integrated output is a ramp that is applied to the VCO, causing the frequency to sweep. Slope of the ramp is determined by the time constant of the integrator and the magnitude of the current. Circuit details are shown in Figure 5.10.

The slew current is inserted at the junction of R_2 and C, not directly into the summing junction of the operational amplifier. If the current were applied directly to the op amp, there would be an output step component (in addition to the desired ramp) of $I_s R_2$ whenever the slew current was turned on or off. The step could cause the loop to jump out of lock, depending on the circuit parameters.

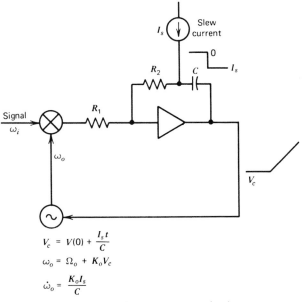

$$V_c = V(0) + \frac{I_s t}{C}$$

$$\omega_o = \Omega_o + K_o V_c$$

$$\dot{\omega}_o = \frac{K_o I_s}{C}$$

Figure 5.10   Frequency sweep circuit.

When the loop locks, the integrator has exactly the right charge needed to hold the VCO at the signal frequency. The loop overcomes the injected slewing current by means of a DC output from the PD, which, in turn, is produced by a dynamic-lag phase error (Chapter 4).

After lock has been achieved, the phase error constitutes a loop stress that impairs tracking capability in the presence of noise or other disturbances. It may be advisable to shut off the slew current once lock has been verified. (Lock detectors are described later in this chapter.) Slew shutoff is particularly necessary if the signal is subject to fast fading; the sweep circuit could carry off the VCO frequency in the event of a fade and the entire sweep range would have to be searched before the signal could be reacquired.

However, the decision to shut off the slew does not have to be particularly fast. The loop does hold lock with the slew applied so a sufficient time can be taken for lock verification to assure a reliable decision.

The simplicity described so far and the freedom to perform a leisurely lock verification is offered only with a closed-loop sweep. One could also perform an open-loop sweep,[15] but it then becomes necessary to detect frequency agreement very rapidly, and then quickly shut off the sweep and close the loop. In principle, the sweep rate is no longer restricted by the ramp tracking limits of the loop, but the need for reliable measurement of frequency coincidence in the presence of noise still places limits on rates.

An ingenious variant on closed-loop search is shown in Figure 5.11. The loop filter doubles as a low-frequency sweep oscillator. A positive-feedback network causes the active filter to oscillate at some low frequency while the

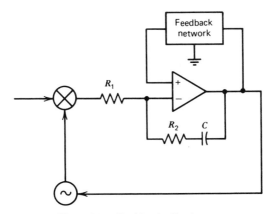

Figure 5.11   Positive-feedback sweep.

loop is out of lock. When the sweep encounters the signal frequency and the loop has locked, there is so much negative feedback around the phaselock loop that the local positive feedback around the op amp is overcome and the oscillation is suppressed, so the loop tracks normally.

The feedback can be through a Wien bridge or a phase-shift network or other similar, broadly selective network. Sinusoidal sweep waveforms are most likely, so use of this circuit entails more search time than does a linear sweep waveform. Amplitude of the sweep is set by the clipping limits of the op amp.

Sweep can also be applied to third-order loops. Since the third-order loop is supposed to be better able to track a frequency ramp, one might expect that a faster sweep should be possible. Unfortunately, the extra complexity of the third-order loop has so far prevented discovery of a practical method of achieving the supposed improvement.

On the contrary, there is fear among designers that acquisition with a closed third-order loop might be unstable, and various expedients to avoid instability are often encountered. One solution is to employ open-loop search, as mentioned earlier. This search requires fast recognition of zero beat and immediate closing of the loop; these are tricky operations although they have been accomplished successfully.

Another solution is to search with a closed second-order loop and then insert the additional loop integrator after lock has been achieved. Search rate cannot be any greater than allowed for the second-order loop. Tausworthe and Crow[9] show that the third-order poles should be overdamped to assure retaining lock through the loop-switching operation.

No recognition has been given to the fact that a third-order loop must acquire *three* variables: phase, frequency, and frequency rate. It may be necessary to engage in a two-dimensional search for both frequency and frequency rate. (Phase presumably is self-acquired.) This subject needs more investigation.

Discriminator-Aided Frequency Acquisition

If the input signal-to-noise ratio is large enough, a frequency discriminator can be used in a conventional automatic frequency control loop to bring the VCO frequency close to that of the signal. Phaselocking occurs when the frequency error is brought within the lock limit.

A typical block diagram and the linearized loop equations are shown in Figure 5.12. The phase loop has little effect when out of lock; the VCO is controlled almost exclusively by the frequency loop. After locking, the phase loop dominates because it has much larger DC gain (infinite, in fact, because of the phase-integrating property of the VCO) and the discriminator can then be disconnected if desired.

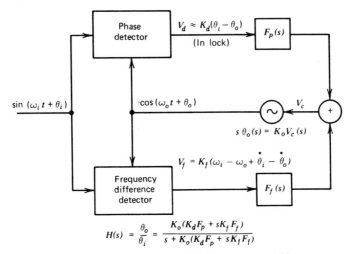

Figure 5.12 Discriminator-aided frequency acquisition.

If a second-order transfer function is an appropriate choice for the PLL, then a first-order loop is appropriate for the frequency loop; that is, the loop filter for the frequency loop would be a simple integrator, without any lead zero. The two loops could share the same operational integrator, as shown in Figure 5.13.

After phaselock has occurred, it can be shown[16] that the overall loop is still of second order and that the frequency path adds to the damping. One could leave the discriminator connected permanently and merely weight the relative contributions of phase and frequency detectors so as to obtain the desired damping.

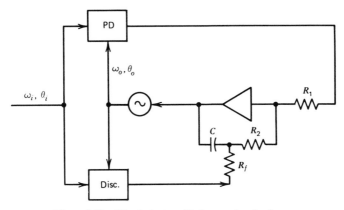

Figure 5.13 Discriminator-aided second-order loop.

A sweep operation can proceed satisfactorily with a poor input signal-to-noise ratio because the PLL is a coherent device and can recover a signal buried in noise. By contrast, a discriminator is an incoherent device and cannot distinguish between signal and noise. Its average output tends to be the average frequency of signal plus noise: approximately the centroid of the total spectrum applied to the discriminator. If noise dominates, the discriminator output is determined almost entirely by the noise properties and the signal is suppressed. A discriminator can be used only under conditions where it provides useful information on signal frequency, which ordinarily means the input signal must exceed the noise. No good analysis exists to provide quantitative insight into the problem. As a rule of thumb, one should be cautious if the input SNR is less than +6 dB and should be very concerned—perhaps to the point of abandoning the discriminator—if input SNR is less than 0 dB.

For a second-order loop, we have shown that pull-in time is proportional to the square of the initial frequency difference and it is reasonable to argue that sweep time is proportional to the first power of frequency difference. If a linear discriminator is used, then it can be shown[16] that the discriminator-aided frequency-acquisition time is proportional to the logarithm of the frequency error. Where applicable, discriminator aiding is the fastest available method of frequency acquisition.

Conventional circuits (e.g., Foster-Seely, pulse averaging, etc.) can be used for the discriminator, but better alternatives exist. Instead of a measurement of absolute frequency, the acquisition discriminator should provide an indication of the frequency difference between the incoming signal and the VCO; a *frequency-difference* discriminator is needed.

Richman[3] describes a frequency-difference discriminator, which he calls the *quadricorrelator*. A block diagram and pertinent equations are shown in Figure 5.14. The input bandpass signal is translated to two quadrature baseband components by the pair of multipliers (mixers, phase detectors) driven by the oscillator. Baseband lowpass filters establish the frequency-difference range over which the circuit will operate. (Richman also included highpass sections in the baseband filters to disconnect the quadricorrelator automatically for very small frequency differences at which the PLL takes control.)

One of the filtered baseband channels is differentiated and then multiplied by the other channel. The product contains a DC component proportional to the frequency difference between signal and oscillator, including the proper sign. It provides an excellent frequency-difference indication. (There is also a sinusoidal ripple component of equal amplitude at double the difference frequency. This can be a serious nuisance if the quadricorrelator were to be used as an FM demodulator, but the difference

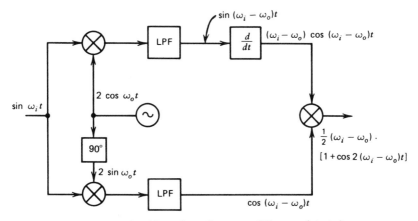

Figure 5.14 Quadricorrelator (frequency-difference detector).

frequency goes to zero when the phase loop locks, so the ripple vanishes when the quadricorrelator is used as an acquisition aid.)

Circuitry for the quadricorrelator might seem rather complex, but appearances are deceptive. One of the input phase detectors is also used as the main phase detector for the associated PLL and the other input phase detector could double as the coherent amplitude detector described shortly. The oscillator, of course, is also part of the PLL. Therefore, only the filters, differentiator, and third multiplier are additional components that are charged to frequency acquisition.

Another form of frequency aid is provided by the popular sequential phase-frequency detector described in Chapter 6. It develops a constant DC output when the loop is out of lock. Amplitude is independent of frequency error, but the sign of the output shows whether the VCO frequency is higher or lower than the signal. The phase detector thereby applies a slew voltage to the loop filter, so the response is more like a sweep than a discriminator. The method is attractive because all the components are built into an inexpensive integrated circuit.

5.4 DIVERSE MATTERS

There are several items, such as lock indicators, variable-bandwidth methods, and loop memories that are more or less associated with the subject of acquisition but do not fit into a neat heading of their own. They are grouped together in this section.

Lock Indicators

A method of lock indication employed almost universally is the *quadrature phase detector*, also known as the *auxiliary phase detector* or the *coherent amplitude detector*, as shown in Figure 5.15. The quadrature phase detector has the received signal applied as one input and a 90°, phase shifted version of the VCO as the other. The main phase detector has an output voltage proportional to $\sin\theta_e$, whereas the quadrature output is proportional to $\cos\theta_e$. In the locked condition θ_e is small, and so $\cos\theta_e \approx 1$. When the loop is unlocked, the outputs from both phase detectors are beat-notes at the difference frequency, and the DC output is almost zero.

Thus the filtered output of the quadrature detector provides a useful indication of lock. The magnitude of the output voltage, relative to that obtained from a noisefree stable input, provides a measure of the quality of lock. (If θ_e jitters, the average of $\cos\theta_e$ is less than unity.) When used in this manner, the smoothed voltage is sometimes known as the "correlation" output.

It is also possible to use the same voltage as a source of coherent AGC control voltage. This topic has been analyzed by Victor and Brockman[17] and is covered further in Chapter 8 of this book.

The output-smoothing filter is a vital part of a practical lock indicator. Without smoothing, the indication will flicker on and off because of noise, giving false indications of lock or loss of lock. If there is excessive smoothing, the lock, or unlock, indication is delayed unduly from the time of its actual occurrence. A compromise amount of smoothing is required. Tausworthe[18] has performed a detailed analysis of the problem and has produced design curves.

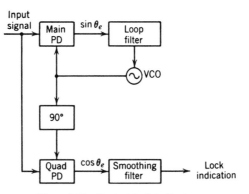

Figure 5.15 Typical lock indicator.

Wide Bandwidth Methods

Speed of acquisition—by pull-in by sweep, or by discriminator aiding— is improved by widening the loop bandwidth. A loop can be built to have a large bandwidth for rapid acquisition and a much narrower bandwidth for good tracking in the presence of noise. It should be apparent that increase of bandwidth can be successful only if signal-to-noise ratio is sufficiently large. If the bandwidth change brings the loop close to threshold, acquisition is not likely.

Bandwidth may be changed by any of several methods. A straightforward approach is to switch loop filter components. (It is usually advisable to switch the resistors only; if a new capacitor were switched in, the integrator charge would be disturbed, and the switching process might cause loss of lock.)

Bandwidth can also be changed by switching the loop gain. Richman[19] has compared filter switching and gain switching and has devised some useful approaches.

The switching command signal can be the lock indication voltage from the quadrature phase detector. When the loop is out of lock, the absence of indication voltage permits the switches to be in their wideband position. When the loop locks, the indication voltage appears and forces the switches into their narrowband position.

If coherent AGC is employed, the same effect can be obtained without switches. In the unlocked condition there is no AGC voltage and the signal level at the phase detector is large. When the loop locks, AGC voltage appears and reduces the applied signal voltage. Since phase detector gain —and therefore, loop gain—is proportional to signal level, the loop bandwidth and damping both decrease automatically when the loop locks; no switches are needed.

Memory

In the absence of disturbance the VCO of a second-order loop tends to remain close to its locked frequency in the event of a signal dropout. When the signal returns, reacquisition by lock-in or pull-in should be very rapid. Thus the loop has a frequency memory. Frequency information is stored in the form of charge in the integrator. When signal drops out, the loop opens and the discharge time constant of the integrator is $|A| R_1 C$. (See Figure 2.2 for nomenclature.) The gain A is unity in a passive loop, so the memory evaporates fairly quickly. However, in an active loop A can be very large, and one would expect long holding times.

This expectation is only partly met in actual equipment. Any real DC amplifier will have some offset and drift, and any real phase detector will have some small DC output (due, for example, to imperfect balance), particularly if there is a noise input. These drifts, unbalances, offsets, and rectified noise all combine to form a small slewing voltage that is integrated and drives the VCO away from its proper frequency.

A first-order loop has a volatile phase memory. Upon signal dropout, the VCO phase immediately begins to drift off from its locked condition at a rate equal to the frequency difference between the signal and the free-running VCO. In other words, the VCO instantly reverts to its free-running frequency when the signal disappears.

Because it has frequency memory, a second-order loop retains its phase information much better than a first-order loop.

A third-order loop has frequency-rate memory, in addition to frequency and phase memories. The third memory can be helpful if input frequency is changing during a signal fade.

REFERENCES

1. A. J. Viterbi, *Acquisition and Tracking Behavior of Phase-Locked Loops*, External Publication No. 673, Jet Propulsion Laboratory, Pasadena, CA, July 1959.

2. A. J. Viterbi, *Principles of Coherent Communication*, McGraw-Hill, New York, 1966, Chap. 3.

3. D. Richman, "Color Carrier Reference Phase Synchronization Accuracy in NTSC Color Television," *Proc. IRE*, Vol. 42, pp. 106–133, January 1954.

4. F. M. Gardner, "Hangup in Phase-Lock Loops," *IEEE Trans.*, COM-25, pp. 1210–1214, October 1977.

5. S. W. Golomb, L. D. Baumert, M. F. Easterling, J. J. Stiffler, and A. J. Viterbi, *Digital Communications with Space Applications*, Prentice-Hall, Englewood Cliffs, NJ, 1964.

6. C. E. Gilchriest," Pseudonoise System Lock-in", *JPL Research Summary No. 36-9*, Vol. I, pp. 51–54, July 1, 1961.

7. U. Mengali, "Acquisition Behavior of Generalized Tracking Systems in the Absence of Noise," *IEEE Trans.*, COM-21, pp. 820–826, July 1973.

8. S. A. Meer, "Analysis of Phase-Locked Loop Acquisition: A Quasi Stationary Approach," *IEEE Conv. Rec.*, Vol. 14, Pt. 7, pp. 85–106, 1966.

9. R. C. Tausworthe and R. B. Crow, *Practical Design of Third-Order Phase-locked Loops*, Report 900-450, Jet Propulsion Laboratory, Pasadena, CA, April 27, 1971.

10. L. J. Greenstein, "Phase-Locked Loop Pull-in Frequency," *IEEE Trans.*, COM-22, pp. 1005–1013, August 1974.

11. K. Hiroshige, "A Simple Technique for Improving the Pull-in Capability of Phase-Lock Loops," *IEEE Trans.*, SET-11, pp. 40–46, March 1965.

12. P. K. Runge, "Phase-Locked Loops with Signal Injection for Increased Pull-in Range and Reduced Output Phase Jitter," *IEEE Trans.*, COM-24, pp. 636–643, June 1976.

13. R. Adler, "A Study of Locking Phenomena in Oscillators," *Proc. IEEE*. Vol. 61, pp. 1380–1385, October 1973.

14. J. P. Frazier and J. Page, "Phase-Lock Loop Frequency Acquisition Study," *IRE Trans.*, SET-8, pp. 210–227, September 1962.

15. A. Blanchard, *Phase-Locked Loops*, Wiley, New York, 1976, Chapter 11.

16. F. M. Gardner, "Acquisition of Phaselock," *Conference Record of the International Conference on Communications*. Vol, I, pp. 10-1 to 10-5, June 1976.

17. W. K. Victor and M. H. Brockman, "The Application of Linear Servo Theory to the Design of AGC Loops," *Proc. IRE*, Vol. 48, pp. 234–238, February 1960.

18. R. C. Tausworthe, "Design of Lock Detectors," *JPL SPS, 37-43*, Vol. III, pp. 71–75, Jet Propulsion Laboratory, Pasadena, CA, January 31, 1967.

19. D. Richman, "DC Quadricorrelator: A Two Mode Sync System," *Proc. IRE*, Vol. 42, pp. 288–299, January 1954.

Chapter Six

Loop Components

An elementary PLL consists of a phase detector, a loop filter and a voltage-controlled oscillator. Design of each component is discussed in this chapter. To give a thorough presentation of phase detectors, it is necessary to introduce the elementary properties of limiters as well.

6.1 LOOP FILTERS

In the early days of phaselock, DC amplifiers were unreliable, costly, and subject to excess drift. Equipment designers were compelled to avoid DC amplifiers and therefore used only passive filters. For this reason, much of the literature implies that a passive filter is somehow the "natural" filter configuration, while the idea of an active filter often gets short shrift.

Yet, it is shown repeatedly in the preceding chapters that a high-gain loop with an active filter will outperform a passive-filter loop in almost all respects. It should be obvious by now that loop filters (for second- or higher-order loops) ought to contain ideal integrators; an operational integrator is a good approximation to ideal, but a passive filter is only a pale imitation. Better performance almost always will be obtained through use of an active filter.

The old prejudice against DC amplifiers is no longer valid. Integrated-circuit, solid-state DC amplifiers of excellent reliability are available at remarkably low cost. Offset and drift of a good op amp are better than can be achieved in the phase detector. An active filter often leads to simpler overall circuitry even though the filter circuit must contain an amplifier.

In short, active filtering should be given consideration for nearly all applications. This book has been written with the tacit assumption that an active filter is the norm and that a passive filter is an inferior substitute that one accepts—along with degradation of performance—only for compelling reasons.

Large DC gain can be obtained using a passive filter cascaded with a DC amplifier; in principle, the same K_v can be achieved in this manner as by using an active filter. Why is an active filter the better choice?

The two filter configurations are shown in Figure 6.1. The same phase detector with gain K_d and the same VCO with gain K_o is used in both configurations. We define the amplifier gains as $A_p(f)$ and $A_a(f)$ for the passive and active filters, respectively. Amplifiers are frequency dependent with 3-dB bandwidths of f_p and f_a. Gains at DC must be identical to have identical $K_v = K_o K_d A$, so $A_a(0) = A_p(0) = A$.

The two configurations differ significantly in the sizes of passive components needed to realize any specified values of natural frequency and damping. By use of (2.11), the design formulas for the lag time constants are:

$$\tau_{1p} = \frac{K_0 K_d A}{\omega_n^2}$$

$$\tau_{1a} = \frac{K_0 K_d}{\omega_n^2} \tag{6.1}$$

It is immediately evident that τ_{1p} is A times as large as τ_{1a}, a factor that can be overwhelming if A is large and bandwidth is small.

Next, we observe that each amplifier introduces an extra pole into the PLL. (We assume that the amplifier response can be approximated by a single-pole model.) The pole corner frequency is at f_p for the passive filter but at $f_a A R_1 / (R_1 + R_2)$ for the active filter. For equal locations of the extra PLL pole, the active filter allows the amplifier pole to be at a much lower frequency. Deleterious effects of extra poles are explained in Chapter 8.

Finally, we note that the amplifier gain in the passive filter has a direct effect on natural frequency and damping, while amplifier gain in the active filter has no influence on these loop parameters. Amplifier gain in the passive loop must be set with some accuracy (e.g., $\pm 10\%$), but in the active loop it is only necessary that the gain exceed some minimum established by requirements on K_v. Gain tolerance is much looser for the active loop.

Circuit precautions must be observed when using active filters. An op amp usually has DC gain of 10^4 to 10^8 or even more, depending on type. There inevitably is DC offset in the circuit, arising mainly from the phase detector, although the amplifier also contributes. When the loop is out of lock, there is no signal-generated DC output from the PD to counteract the offset. Therefore, the offset is integrated until, eventually, the amplifier is driven to its saturation limit.

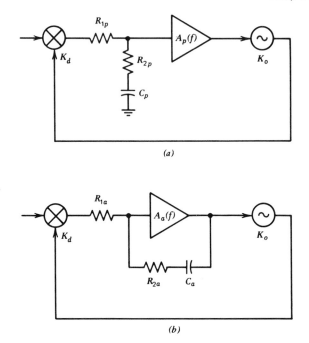

Figure 6.1 Loop filter comparison: (*a*) passive filter with DC amplifier. (*b*) active filter.

Amplifier characteristics change at saturation; appearance of a signal at the loop input may not be sufficient to bring the amplifier out of saturation. In other words, self-acquisition—frequency pull-in—can be impaired by saturation.

An aided acquisition technique, such as frequency sweep or discriminator assistance, prevents saturation. Sweep circuits are designed to reverse direction at the sweep limits, which are chosen within the amplifier saturation limits. The aiding voltage or slew current must be sufficient to overcome any circuit offset. Avoidance of saturation is yet another reason to use aided frequency acquisition instead of pull-in.

If a particular application absolutely prohibits aided frequency acquisition, then saturation can be avoided by providing DC feedback around the amplifier. The product of closed-loop DC gain and equivalent input offset must be less than the amplifier clipping level. Choice of the resulting DC gain is a compromise between degradation of PLL tracking performance and the initial frequency error caused by amplified DC offset.

Another, more subtle, difficulty can arise because the typical op amp is not capable of handling high-frequency signals of large amplitude. Instead, it goes into slew limiting. This may not cause a problem under locked conditions, but the op amp must also cope with the beat-note that is present when the loop is still out of lock.

If slew limiting were symmetrical on positive and negative half-cycles, there still would be no difficulty. However, slew limiting is not symmetrical; there is, in effect, some rectification of the beat-note. The resulting DC component could aid frequency acquisition or oppose it with equal probability. To avoid acquisition problems, the op amp selected should avoid slew limiting for any beat-note that might be encountered.

A related effect arises from the ripple output of the phase detector. In Chapter 3 we identify a PD output at double the signal frequency and then ignore it in the noise performance analysis. Other types of phase detectors can produce more severe forms of ripple. Whatever the form of the ripple, it can rarely be ignored in real hardware.

Specifically, a low-frequency op amp cannot tolerate much high-frequency or square-wave ripple at its input. Some measures must be taken to suppress the ripple before it reaches the amplifier. Several low-ripple phase detectors are identified later in this chapter. Another expedient is to place ripple filters between the phase detector and the op amp. Simple, lowpass filters are used in some applications, while notch filters have been used in others.

6.2 VOLTAGE-CONTROLLED OSCILLATORS

There are many requirements placed on VCOs in different applications. These requirements are usually in conflict with one another, and therefore a compromise is needed. Some of the more important requirements include the following:

1. Phase stability (Spectral purity).
2. Large electrical tuning range.
3. Linearity of frequency versus control voltage.
4. Large gain factor (K_o).
5. Capability for accepting wideband modulation.
6. Low cost.

The requirement for phase stability is in direct opposition to all the other five requirements. To obtain any of the wideband features we must inevitably sacrifice phase stability.

Oscillator Types

The four types of VCO in common use are given below in order of
decreasing stability:

1. Crystal oscillators (VCXO).
2. Resonator (LC, coaxial, or cavity) oscillators.
3. RC multivibrators.
4. YIG-tuned oscillators.

In today's technology the stablest oscillators are those using high-Q,
vacuum-mounted, 2.5- or 5.0-MHz, fifth-overtone, AT-cut crystals. Reference 1 provides a review and a large bibliography.

A circuit commonly used (Figure 6.2a) is a variation on the familiar
Pierce crystal oscillator.[2,3] The crystal is operated as an inductance, and
capacitors C_1 and C_2 adjust the amount of feedback. A varactor diode
provides a small variation of C_2 and causes a pulling of the oscillation
frequency.

The tuning range of this circuit is very small when high-Q crystals are
used. To obtain a greater range it is common practice to use ordinary
AT-cut crystals in their fundamental series-resonant mode* in the circuits
of Figures 6.2b and 6.2c. The varactor is in series with the crystal and
effectively varies the resonant frequency over a greater range than is
possible in the Pierce circuit. Other oscillator circuits are described in
Ref. 4.

Figure 6.2c shows some circuit details that contribute to good performance and are applicable to other oscillator configurations.

- The varactors are connected in series opposing. Therefore, if the RF
  voltage should exceed the DC bias, forward current will be blocked by
  whichever diode is still reverse biased.
- A series inductance L_s is used to tune the varactor capacitance to series
  resonance at the crystal frequency. This inductance permits operation
  both above and below the crystal resonant frequency. (Without the
  inductance, the varactor can only pull the crystal to higher frequencies,
  where the crystal reactance is inductive.)
- A parallel inductance L_x tunes out the parallel capacitance of the crystal.
  Analysis of the equivalent circuit of the crystal and its tuning elements
  demonstrates that this measure increases the tuning range of the oscillator.

*Overtone crystals have a narrower pulling range than fundamental crystals.

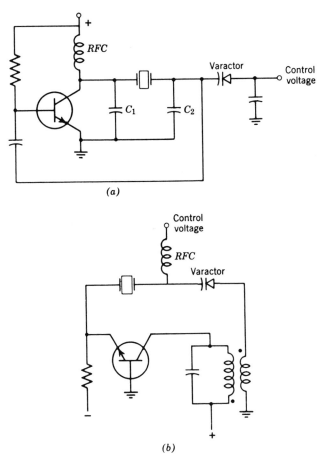

(a)

(b)

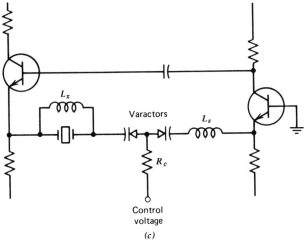

(c)

Figure 6.2 VCXO circuits: (a) modified Pierce oscillator; (b) grounded-base oscillator; (c) emitter-coupled oscillator.

97

- Reactances in the crystal and *LC* series network tend to be large. To avoid dissipative losses, the isolation resistor R_c must also be large. Better isolation may be attainable from RF chokes.

Speed of response of oscillation frequency to a changing control voltage is often thought to be restricted by the bandwidth of the crystal; this is incorrect. In fact, frequency will change just as quickly as the capacitance of the varactor can be changed. This change depends only on the control voltage actually applied to the varactor, a voltage that is response restricted by the *RC* lowpass filter composed of the isolation resistor and the varactor capacitance. Fast response of a VCXO—much faster than is characteristic of the crystal bandwidth—is entirely feasible if proper care is taken in the drive circuits.

Phase-stability is enhanced by a number of factors:

1. High Q in the crystal and circuit.
2. Low noise in the amplifier portion.
3. Temperature stability.
4. Mechanical stability.

The precision 5-MHz crystals mentioned have an unloaded Q of approximately 2×10^6. Other crystals can be expected to have unloaded Q's in the range of 10,000 to 200,000. Circuit losses inevitably degrade the intrinsic Q of the crystal alone; these losses must be minimized for best performance. In a series-mode crystal the driving and load impedances should be as small as possible to avoid degradation of Q.

Much of the phase jitter of an oscillator arises from noise in the associated amplifier. The transistor (or other device) should be operated in a low-noise condition and, of course, a low-noise transistor should be used. High-frequency thermal and shot noise contribute significantly to the jitter but flicker $(1/f)$ noise is predominant in many applications. Treatment of flicker noise is deferred to later pages.

To obtain a good signal-to-noise ratio in the oscillator (and therefore low jitter) it seems reasonable to operate the circuit at a high RF power level. There is a competing effect, however; excessive vibration of the crystal drives it into nonlinear modes of mechanical damping and the Q is thereby reduced. As a result, there is an optimum drive level for any crystal. Powers of 10 to 500 μW are typical; these levels are usually much smaller than the maximum rated power that is established on heat dissipation limitations.

Crystal parameters are temperature sensitive; to obtain best phase stability, temperature transients and fluctuations should be avoided.

A quartz crystal is an accelerometer; vibration, shock, or even changes of attitude in a gravitational field will cause frequency changes.

When wide tuning range becomes more important than stability, other oscillator types must be used. Rumor tells that X-cut crystals in parallel-mode circuits have been employed in very wide range VCXOs, but as far as is known extreme tuning limits of 0.25 to 0.5% of oscillator frequency are all that has been achieved.

If a wider range is needed, an LC oscillator must be used. In this application, the standard Hartley, Colpitts, and Clapp circuits make their appearance. Tuning may be accomplished by means of a varactor, although saturable inductors have also been used.

When stability is of little importance, large tuning range is needed, and when low cost is a factor, relaxation oscillators such as multivibrators are used. The upper operating frequency of practical relaxation oscillators has been limited to a few tens of megahertz. Linearity of frequency versus control voltage (or current) can be excellent. Multivibrators are available at very low cost as packaged integrated circuits (IC). Stability and linearity of the IC versions have generally been inferior to discrete-circuit multivibrators.

A multivibrator contains a rundown circuit (i.e., a capacitor driven by a controllable current), a voltage pickoff device, and a switch. Frequency of operation is established by the time needed for the capacitor voltage to run down from an initial condition set after switching to the trigger point set by the pickoff device. Phase jitter is caused largely by noise displacement of the pickoff threshold.

Time displacement caused by noise is inversely proportional to the slope of the rundown voltage; a high-frequency multivibrator has less time jitter than a low-frequency unit employing exactly the same circuit except for timing capacitor. Therefore, to obtain improved phase jitter, it is useful to operate a multivibrator at a frequency Nf_o, where f_o is the desired output frequency, and divide the output by N. Time jitter is preserved in a frequency divider circuit, but phase jitter is reduced by a factor of N. This scheme improves the phase jitter by a factor of N over that obtained from a multivibrator operating directly at f_o.

At microwave frequencies, YIG-tuned Gunn oscillators have become popular. They are capable of very large tuning ranges, have highly linear tuning characteristics, and provide useful output power. They are very noisy compared to crystal or LC oscillators, so relatively large bandwidths must be used to lock them. (See next section for relation between bandwidth and residual noise.)

Tuning of the YIG sphere is accomplished by altering a magnetic field; magnetic devices are notoriously slow to respond. A large coil is used to establish the bulk of the magnetic field and a small "tickler" coil is used to accommodate fast input modulation. The large coil cannot change flux rapidly, and the small coil cannot change frequency very far, by comparison to the large coil; but, between them, they provide the large range and fast response that is needed.

Oscillator Phase Noise

Oscillator noise—also called short-term instability or phase jitter—is an extensive topic. Discussion of the problem and further bibliographies may be found in Refs. 5 and 6. Attention here is restricted to the effects of noise on phase-error fluctuations in a phaselocked loop, a very small portion of the subject.

Let us consider the loop model of Figure 6.3. The oscillator is taken to be a perfect, jitter-free oscillator with a phase output of ϕ_o, followed by an internal disturbance source that adds a phase jitter ϕ_n. Phase delivered to the external oscillator output terminal is $\theta_o = \phi_o + \phi_n$.

Phase of the input signal is θ_i and the phase error in the closed loop is $\theta_p = \theta_i - \theta_o$. If the input signal is noisefree, then the tracking error can be shown (by the methods of Chapter 2) to be

$$\theta_p(s) = [1 - H(s)][\theta_i(s) - \phi_n(s)] \tag{6.2}$$

Phase noise on the incoming signal and on the local VCO affect tracking error in the same manner (except for sign, which is immaterial for our purposes). The development here is in terms of ϕ_n only, but identical results apply to fluctuating θ_i.

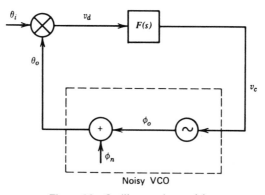

Figure 6.3   Oscillator noise model.

Strictly speaking, a random function such as ϕ_n does not have a Laplace transform, so (6.2) is merely symbolic. More correctly, the spectrum of the phase error is

$$\Phi_p(\omega) = |1 - H(\omega)|^2 \Phi_o(\omega) \qquad \text{rad}^2/\text{Hz} \qquad (6.3)$$

and the variance of the phase error is

$$\overline{\theta_p^2} = \frac{1}{2\pi} \int_0^\infty |1 - H(\omega)|^2 \Phi_o(\omega) d\omega \qquad \text{rad}^2 \qquad (6.4)$$

where $\Phi_o(\omega)$ is the one-sided spectral density of the phase noise ϕ_n in units of rad^2/Hz.

For various reasons, it is more common to consider the spectrum of the frequency fluctuations $\Phi_\omega(\omega)$ instead of phase spectrum. The two spectra are formally related by $\Phi_\omega(\omega) = \omega^2 \Phi_o(\omega)$, where the units of Φ_ω are $(\text{rad}/\text{sec})^2/\text{Hz}$.

Oscillators are afflicted by disturbances of various kinds; these are described presently. The loop tries to track out the disturbances. To minimize the tracking error, the factor $|1 - H(\omega)|^2$ should be as small as possible, which means the loop bandwidth should be as large as possible. This is an intuitively logical conclusion; the loop is better able to track disturbances if it has a large bandwidth.

Loop error is the untracked disturbance. To calculate variance of the error, it is only necessary to substitute expressions for loop transfer function and disturbance spectral density into (6.4).

Typical forms of oscillator noise spectra are shown in Figure 6.4. In addition to the exhibited continuous spectrum, there can also be discrete spectral components arising from power-supply hum or interference from other signals.

Starting at the high-frequency end of the noise spectrum disposes of the simplest explanations first. Any physical circuit has a high-frequency band limit imposed by parasitic effects. Noise from any source falls off in the cutoff region, assuring a finite noise-variance contribution from high frequencies.

Noise in the next-lower region originates from white noise, primarily in the amplifiers and output circuits following the oscillator. It results in a flat phase-noise spectrum that is equivalent to a parabolically rising frequency-noise spectrum. The source of this disturbance is simple, additive noise of the familiar variety.

Phase flicker appears at somewhat lower frequencies. It is distinguished by a phase-noise spectrum that varies approximately as $1/f$. Phase flicker

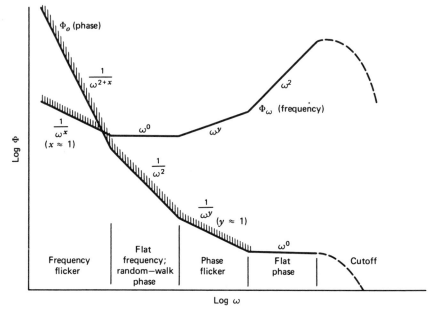

Figure 6.4 Oscillator noise spectra. Hachure marks indicate regions of nonstationary or nonexistent autocorrelation.

originates from flicker noise intermodulation in nonlinear circuits, such as frequency multipliers,[7] that are outside the oscillator feedback loop.

At this point we begin to encounter mathematical difficulties. There exists no Fourier transform of the $1/f$ phase flicker spectrum, so the autocorrelation function does not exist. Since spectral density is formally defined as the Fourier transform of the autocorrelation, the meaning of the "spectrum" is not clear. The same problem will plague us as we examine lower-frequency portions of the disturbances. Suffice it to say that laboratory instruments called "spectrum analyzers" give outputs consistent with Figure 6.4 and that formally using the questionable spectra in (6.4) gives no difficulty in most cases. [For the phase flicker spectrum it is necessary to invoke high-frequency cutoff to arrive at finite jitter from (6.4).]

Spectra whose existence is questionable are indicated by hachure marks in Figure 6.4.

Moving one step lower in frequency, we next encounter flat frequency noise that could be generated, for example, by white noise added into the control terminal of the VCO. Phase is the integral of frequency; the integral of white noise is a sort of random walk and is not stationary. Since we have defined spectral density only for stationary processes (Appendix

A), the phase spectrum of flat frequency noise is not defined properly. However, using the formal relation $\Phi_o = \Phi_\omega/\omega^2$ does not lead to difficulty in the PLL with this form, so we overlook the mathematical peculiarities.

At the lowest frequency range—closest to the oscillator carrier frequency—lies frequency flicker noise. It has been the source of much puzzlement and is still rather mysterious.

Its existence has been known for many years, but its causes are unclear. Thermal and shot noise are explained very well as the random motion of electrons, but there is no similar explanation of flicker noise.

Flicker noise arises as a low-frequency disturbance in almost all active electronic devices, but there is no widely accepted explanation of the mechanism that transfers the low-frequency noise into RF phase fluctuations. (There is even significant flicker noise in a quartz crystal itself,[12] but that does not explain frequency flicker in other types of oscillators.)

The frequency flicker noise that appears to be present in all oscillators causes the frequency noise spectrum to have the form

$$\Phi_\omega(\omega) = N_1\left(\frac{\omega_o}{\omega}\right)^x \tag{6.5}$$

where N_1 is the spectral density in $(rad/sec)^2/Hz$, measured at a frequency ω_o. Exponent x is a number in the vicinity of 1.

The spectrum (6.5) is very troublesome mathematically. First of all, its Fourier transform does not exist, so flicker noise does not have an autocorrelation function. Next, variance of a quantity is given by the integral of its spectrum over all frequencies from zero to infinity. The integral of (6.5) is improper at both limits for some values of x. For any x less than 2, the integral diverges at high frequencies. However, that problem can be accommodated by imposing the high-frequency cutoff; no circuit can have infinite bandwidth. Also, the PLL responds to the phase jitter, whose spectrum is $1/\omega^2$ times that of the frequency spectrum. Therefore, the upper limit does not cause any trouble in analysis of a phaselock loop, even without an upper cutoff.

If $x \geqslant 1$ then the integral of (6.5) diverges at the zero-frequency limit also. This is much more troublesome than high-frequency divergence. A low-frequency cutoff might be imposed to obtain tractable analysis, but where should a cutoff frequency be placed? Measurements[8] show the $1/f^x$ slope continuing to frequencies at least as low as 1 cycle/month. There is no evidence that a cutoff exists in nature.

How does a PLL respond to frequency flicker in its oscillator? Inserting a frequency flicker spectrum into (6.4) and remembering that $\Phi_o(\omega) = \Phi_\omega(\omega)/\omega^2$, we arrive at infinite phase-error variance for any PLL with finite K_v. In other words, when our model is applied to a real PLL, the analysis predicts that the loop cannot remain locked if the VCO has flicker noise.

We are faced with a paradox. On the one hand, measurements on oscillators show flicker noise very clearly and almost universally. On the other hand, experience tells us that PLLs lock very well and are only mildly perturbed by oscillator noise.

Something is wrong with the model, or with the interpretation thereof. It is hoped that the mathematical difficulty can be resolved in the future. In the meantime, how is the effect of flicker noise to be treated in PLLs that must be built now?

Analysis of the first-order loop appears to be hopeless with the present model. Existing analyses[9-11] lead either to an infinite variance or to a variance that grows logarithmically with duration of observation time. Experience does not confirm these predictions, although it is conceivable that the growth rate is so small as to escape notice.

An engineering solution is feasible for a second- or higher-order loop. If the loop filter contains a perfect integrator, (6.4) has a finite result, even when confronted with a flicker spectrum. Good prediction of experimental observations is obtained if the integrator is assumed to be perfect.

Assuming a second-order loop with perfect integrator, and assuming $x = 1$, we calculate phase jitter from (6.4) and (2.12) to be

$$\overline{\theta_p^2} = \frac{N_1 \omega_o}{4\pi \omega_n^2} f(\zeta) = \frac{N_1 \omega_o}{4\pi (2B_L)^2} g(\zeta) \tag{6.6}$$

where $g(\zeta) = (\zeta + 1/4\zeta)^2 f(\zeta)$ (by using Table 3.1) and

$$f(\zeta) = \frac{1}{4\zeta\sqrt{\zeta^2 - 1}} \ln \frac{2\zeta^2 - 1 + 2\zeta\sqrt{\zeta^2 - 1}}{2\zeta^2 - 1 - 2\zeta\sqrt{\zeta^2 - 1}} \qquad (\zeta > 1)$$

$$= \frac{1}{2\zeta\sqrt{1 - \zeta^2}} \left[\frac{\pi}{2} - tan^{-1} \left(\frac{2\zeta^2 - 1}{2\zeta\sqrt{1 - \zeta}} \right) \right] \qquad (\zeta < 1) \tag{6.7}$$

$$= 1 \qquad (\zeta = 1)$$

The functions $f(\zeta)$ and $g(\zeta)$ are plotted in Figure 6.5. For a loop of specified noise bandwidth B_L, the minimum flicker jitter is obtained for a slightly overdamped condition ($\zeta = 1.14$). These results coincide with those obtained by Gray and Tausworthe.[9]

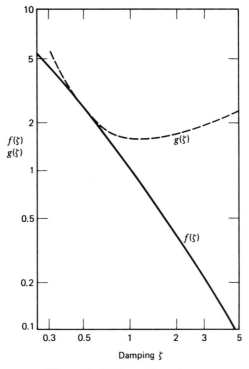

Figure 6.5 Flicker-variance factors.

Equation 6.6 shows variance to be inversely proportional to the square of bandwidth. If bandwidth is made too small, the phase error exceeds the linear limits that implicitly underlie all of this analysis and the results become inaccurate. Further narrowing of the bandwidth causes the loop to lose lock. One simple measure of oscillator stability is the narrowest loop bandwidth at which lock still holds.

The inverse-square dependence of variance on bandwidth is a consequence of assuming $x = 1$. If x departs somewhat from unity, the exponent of bandwidth dependence departs somewhat from -2.

Low-Flicker Design

Inasmuch as the origin of flicker noise is not properly understood, its alleviation is more an art than a science. The following paragraphs furnish a brief synopsis of the state of the art.

All components should exhibit low noise; the most critical components are the active devices used to provide gain for positive feedback in the oscillator. One might think that low-noise-figure, high-frequency transistors would be most appropriate but this is an incorrect assumption. Small-junction devices exhibit greater flicker noise than devices with junctions of larger area; however, the reasons are obscure. A high-frequency power transistor with large emitter area might be the best choice of bipolar transistor.

There is also evidence that junction field-effect transistors are better since they tend to have lower flicker noise than bipolar devices. Insulated-gate field transistors are much poorer and should be avoided.

At the time of this writing, discrete devices (both bipolar and FET) have lower noise than integrated circuits.

The essential elements of a feedback oscillator consist of a selective network and a sustaining amplifier. In simple oscillators, the oscillation level is established by signal limiting in the amplifier. Impedances of a limiting amplifier are indeterminate and time varying, a condition that can adversely affect the selective network. Stability is improved by separating the sustaining amplifier from the limiting function so that only constant, stable, low-loss impedances are presented to the selective network.[4]

Fluctuations of the parameters of the sustaining amplifier cause frequency instability, even when operated in a linear manner. Negative feedback has a stabilizing effect.[7]

Flicker noise in a transistor is concentrated in the audio and subaudio frequency range. How do such low-frequency disturbances get transferred to sidebands of the RF oscillator frequency? There must be intermodulation between the low-frequency flicker noise and the high-frequency oscillation to produce the sidebands. Second-order nonlinearities in the oscillator would be particularly conducive to intermodulation.[13]

To avoid intermodulation, a linear sustaining amplifier is needed. If this reasoning is correct, an oscillator with automatic level control, rather than limiting, ought to exhibit less flicker noise.

If the level is set by limiting (for convenience and simplicity of the circuit), then one would expect least flicker noise intermodulation from a symmetrical clipper, which has only odd-order nonlinearity. In any clipper the AM-to-PM conversion factor must be small.

6.3 PHASE DETECTORS

Two broad categories of phase detectors can be distinguished: *multiplier* circuits and *sequential* circuits. Multipliers generate the useful DC error output as the average product of the input-signal waveform times the local-oscillator waveform. Operation of an ideal multiplier is described in Chapter 3. Multipliers are zero memory devices. A properly designed multiplier is capable of operation on an input signal buried deeply in noise.

A sequential phase detector generates an output voltage that is a function of the time interval between a zero crossing on the signal and a zero crossing on the VCO waveform. Other details of the waveform do not contribute to the output.

Sequential phase detectors contain memory of past crossing events. They can generate PD characteristics that are difficult or impossible to obtain with multiplier circuits. Because a sequential circuit operates on waveform edges, it can be intolerant of missing or extra crossings; in consequence, it often has poorer noise-handling capability than a multiplier.

Sequential PDs are usually built up from digital circuits (flip flops, gates) and operate with binary, rectangular input waveforms. Accordingly, they are often called "digital" phase detectors and the loops they are used in are often called "digital" phaselock loops. This terminology is incorrect; the output of a sequential PD is an analog quantity and the loops are analog loops. Examples of true digital loops are given later to illustrate correct usage of the term "digital."

Multipliers

If both inputs to an ideal multiplier are sinusoidal, the useful DC output is proportional to the product of the amplitudes of the two inputs and to the cosine* of the phase difference between them. The multiplier produces the scalar product of the two input phasors (see Chapter 3 for analysis).

In addition, there is an unwanted, sinusoidal *ripple* at double the input frequency and with amplitude equal to the maximum available DC output level. Ripple must be suppressed to prevent unwanted sidebands from appearing on the VCO.

Multiplication can be implemented physically by means of a four-quadrant analog multiplier. Such devices are available as monolithic integrated circuits or as encapsulated packages or can be built from discrete devices. Good performance can be obtained, although usable operating frequencies tend to be low and cost tends to be high.

A true multiplier provides a useful analytical model for a phase detector, but it is rarely found in actual equipment. Instead, a *switching phase detector* is far more popular.

---

*The phase *difference* is 90° when the phase *error* is zero.

Let us suppose that the sinusoidal VCO drive to a multiplier phase detector is replaced by a square wave of the form

$$v_o(t) = \text{sgn}[\cos(\omega_i t + \theta_o)] \qquad (6.8)$$

where the signum function is defined as

$$\text{sgn}(x) = 1, \quad x > 0$$
$$= -1, \quad x < 0 \qquad (6.9)$$

The square wave is periodic and can be expanded in a Fourier series as

$$v_o(t) = \frac{4}{\pi}\left[\cos(\omega_i t + \theta_o) - \frac{1}{3}\cos 3(\omega_i t + \theta_o) + \frac{1}{5}\cos 5(\omega_i t + \theta_o) + \ldots\right]$$

$$(6.10)$$

Output of the multiplier is the sum of each individual term of the Fourier series multiplied by the input signal.

Very often the input signal and noise are bandlimited to a narrow spectrum around the carrier frequency; no harmonics are present at the input. In this case, it is easy to show that the only multiplier product containing a low-frequency (near-DC) component is the one associated with the fundamental frequency of the square wave. All other products only contribute high-frequency ripple.

We let the input signal be $v_i(t) = V_s \sin(\omega_i t + \theta_i)$. The average value (DC component) of the product $v_i v_o$, where v_o is given in (6.10), is

$$v_d = \frac{2}{\pi} V_s \sin(\theta_i - \theta_o) \qquad (6.11)$$

Using the notation of Chapters 2 and 3, $K_d = 2V_s/\pi$. In other words, the useful output is identical to that which would have been obtained if the VCO drive were sinusoidal, with amplitude $4/\pi$. The circuit produces exactly the same DC signal and exactly the same low-frequency noise as the equivalent phase detector with sinusoidal drive.

But multiplication by a unit-amplitude square wave is equivalent to periodic switching of the polarity of the input; the multiplier can be replaced, without penalty (except for ripple waveform), by a polarity switch. Since a switch is much simpler and less expensive to build than a linear multiplier, the most common multiplier type of phase detector is really a switching device.

The foregoing describes a full-wave switching PD; it generates output on both halves of the switching cycle. Half-wave circuits gate the input signal

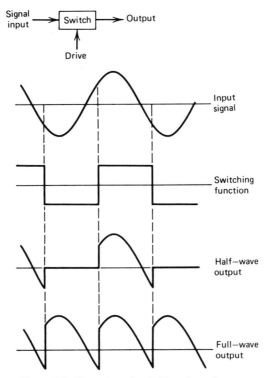

Figure 6.6 Operation of switching phase detector.

$v_i(t)$ on, say, the positive halves of the switching cycle and produce zero output on the negative halves. Waveforms of half-wave and full-wave PDs are shown in Figure 6.6.

Average output of a half-wave PD is

$$v_d = \frac{V_s}{2\pi} \int_{-\theta_o - \pi/2}^{-\theta_o + \pi/2} \sin(\omega_i t + \theta_i)\, d(\omega_i t)$$

$$= \frac{V_s}{\pi} \sin(\theta_i - \theta_o) \tag{6.12}$$

which is exactly half the output obtained from a full-wave PD.

Output amplitude from a switching PD is proportional to the input signal amplitude but independent of the amplitude of the switching voltage.

Examination of the half-wave waveforms of Figure 6.6 reveals that the fundamental ripple frequency is now at the signal frequency, and the nonsinusoidal waveforms change with alterations in phase error; ripple is more difficult to suppress. One measure to reduce ripple is to employ full-wave circuits rather than half-wave circuits, as shown at the bottom of Figure 6.6.

Examples of switching PD circuits are given in Figure 6.7. Many other circuits and devices can be used. The examples show only bipolar transistors and diodes; it is also possible to use field-effect transistors (both junction and insulated-gate transistors), mechanical switches, or optoelectronic devices.

A number of low-cost, prepackaged circuits make excellent phase detectors. One example is the integrated-circuit *balanced modulator* as shown in Figure 6.8. Both single-balanced and double-balanced configurations are used.

Figure 6.9 shows waveforms in the single-balanced circuit of Figure 6.8. In addition to the ripple typical of a full-wave phase detector, there is also a square wave of peak amplitude I_E at the signal frequency, which is much worse than the normal ripple. The square-wave portion of the ripple must be filtered before it enters the output amplifier.

A double-balanced circuit is equivalent to two single-balanced circuits connected together. Polarity of connections is such that the collector-current gaps evident in Figure 6.9 are filled in and the ripple becomes identical to that of a full-wave switching PD, as in Figure 6.6.

Phase-detector gain of the single-balanced circuit is

$$V_d = \frac{2V_s}{\pi R_E} \sin\left(\theta_i - \theta_o\right) \frac{R_B R_C}{R_A + R_C} \tag{6.13}$$

while that of the double-balanced circuit is exactly twice as great. (Equation 6.13 was obtained under the assumptions that the input signal plus noise does not overload the input transistors; that current gain h_{fe} of all transistors is very large; that internal emitter resistance is included in R_E; and that symmetrically located circuit components are perfectly matched.)

The balanced-modulator PDs have differential, balanced outputs with a common-mode DC offset. Most loop-filter circuits require a single-ended input with zero offset. Figure 6.8 shows the differential-to-single-ended conversion performed by a separate operational amplifier. This function can also be accomplished by a current mirror on the same chip.[14]

Prior to the advent of well-balanced integrated circuits, use of an active phase detector was precluded by severe problems of DC offsets. The circuits of Figure 6.8 could never be successful using discrete components. Excellent matching of like components on a single IC chip achieves balances that are unimaginable with separate active components. Even so,

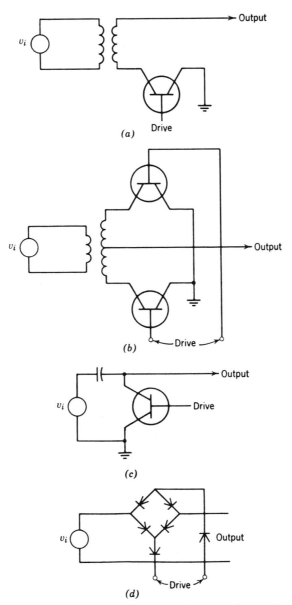

Figure 6.7 Switching phase detector circuits: (a) half-wave series transistor; (b) full-wave transistor; (c) shunt transistor; (d) diode quad.

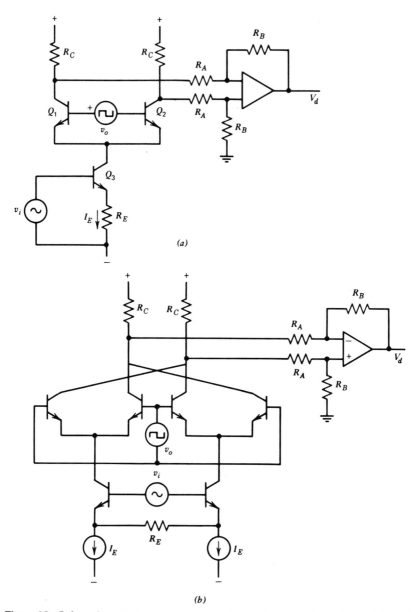

Figure 6.8 Balanced-modulator phase detectors: (a) single balanced; (b) double balanced.

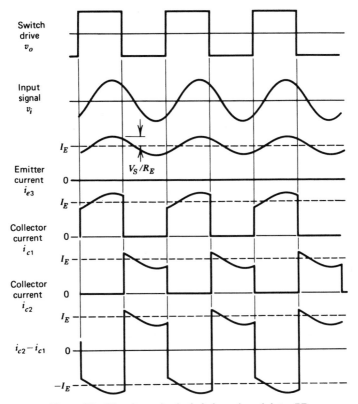

Figure 6.9 Waveforms in single-balanced-modulator PD.

great care must be taken to achieve close balance between external
components and to have low-impedance, well-balanced switching drive if
DC offset is to be held to small values.

Another popular circuit is the diode ring of Figure 6.10. These units are
sold in large quantities at low cost under the name of *double-balanced
mixers*. They have wide bandwidths, are available over an extremely large
frequency range (they operate well at frequencies far above the capabilities
of transistor PDs), impose little burden on the designer, and provide good
performance.

Accurate analysis is difficult. If the diodes are assumed to be ideal and if
the signal voltage is much smaller than the switching voltage, then operation is closely the same as any full-wave switching PD.[15] These conditions
are often violated, so the existing analyses are approximations.

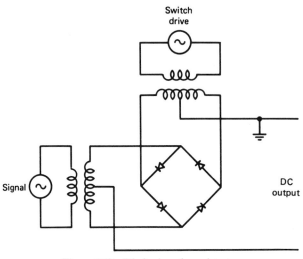

Figure 6.10   Diode-ring phase detector.

Provided that the signal voltage is substantially smaller than the switching voltage, the output characteristic takes the form

$$V_d = V_m \sin(\theta_i - \theta_o) \qquad (6.14)$$

where V_m is proportional to the signal amplitude. (If signal and switching amplitudes are near equal, the PD characteristic becomes triangular[15] instead of sinusoidal.) Typical values of V_m range up to 0.3 to 0.4 V. Since most VCOs require much larger control voltages, a DC amplifier is almost always needed in the loop filter; the diode ring did not become feasible as a PD until good DC amplifiers were developed.

A standard ring is usually specified for 5 mW of sinusoidal drive from a 50-ohm source. Since the diodes are a nonlinear load, and since the time-averaged load is not necessarily matched, the specification is the available power and not the actual power delivered. A DC offset on the order of 1 mV is typical.

If the signal is immersed in noise, the total signal plus noise must be well below the switching drive level if clipping is to be avoided. "High level" circuits, in which two or more diodes are connected in series in each arm of the ring, can accept larger switching drive and therefore larger input signals. Maximum-possible output voltage V_m should be proportional to the number of series diodes.

The diode ring is not really characterized very well for phase-detector service. It is fortunate that the circuit is tolerant of a fairly wide range of operating conditions.

A PD circuit that was once very common—it was considered *the* phase-detector circuit—is shown in Figure 6.11. The hybrid transformer forms the vector sum and vector difference of the two input signals; these are converted to DC signals by the diode rectifiers. The useful output is the difference between the two rectified voltages.

Analysis[16,17] shows that the output is proportional to the sine of the phase error and is a function of the two input amplitudes. If $V_o \gg V_s$, then V_d is proportional only to V_s and is independent of V_o. This feature is found in many different PD circuits, including the diode ring presented above.

Ripple is reduced from that encountered in switching PDs because of the nonlinear filtering action in the RC loads of the peak detectors.

Popularity of this circuit has declined with the advent of good ICs and the packaged rings. Since output is a small difference between two large DC voltages, balance is a critical adjustment if DC offset is to be avoided. It is much easier to buy a well-balanced circuit than to build one from discrete components.

However, the basic circuit should not be dismissed entirely. It has the potential of operating over a frequency range from audio to light. The "transformer" could be realized with a coaxial hybrid junction, or a waveguide magic-T, or even an optical device. Detectors need not be diode rectifiers; they could also be bolometers, thermocouples, or photodiodes. There is still a niche for the circuit at frequencies above the capability of diode rings.

Sample-and-hold phase detectors are sometimes encountered.[18] A sampler is merely a switch that is driven by a short pulse. Signal value at the instant of the pulse is stored on a capacitor until the next sample is taken. If the signal is sinusoidal, the PD characteristic is also sinusoidal, with maximum DC output equal to the peak signal amplitude.

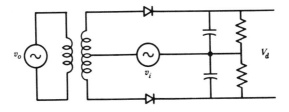

Figure 6.11 Diode rectifier phase detector.

Sample-and-hold PDs are used to lock to harmonics of the sampling rate, to suppress ripple, or in applications where the signal appears in short bursts for which "gated phase detection" is better terminology. Harmonic operation is discussed later.

If noise is absent, and if the input signal does not carry angle modulation, then the sampling always occurs at the same point of the input waveform from one cycle to the next. The DC value (near zero for equilibrium tracking) does not change. Except for possible sharp spikes at sampling times, due to switch imperfections, the voltage on the storage capacitor remains constant; ripple is suppressed completely.

Analysis of a sampled loop is not accomplished readily by the L-transform methods of Chapter 2; it is necessary to use z-transforms instead. Response of a sampled loop, and its stability, differs from the continuous-loop behavior presented in this book.[19, 20]

Nonsinusoidal PD Characteristics

We have examined many different PD circuits and in each case have arrived at a sinusoidal PD characteristic (DC error voltage versus phase error). One might think that a sinusoid is a common property of the different circuits.

Actually, the shape of the characteristic depends on the applied waveforms, not necessarily on the circuit. If rectangular waveforms are applied at both inputs of any multiplier or switching PD, the output characteristic becomes triangular. This result comes out of exactly the same circuit that produces a sinusoidal characteristic when presented with a sinusoidal input.

If waveforms are rectangular, digital circuits can be used in place of analog circuits. The digital-circuit equivalent of a multiplier phase detector is an exclusive-OR gate. Average DC output is a triangular function of the phase error, while the ripple waveform is rectangular with a duty cycle that depends on phase error. Note that the output is an analog quantity despite the fact that a digital circuit and digital input waveforms are used.

The DC characteristic of a sampled PD is exactly the waveform of the input signal. Almost any desired characteristic can be obtained by appropriate shaping of the input. For example, a rectangular PD characteristic occurs if the signal has a rectangular waveshape.

(A rectangular PD characteristic has infinite slope at zero phase error, which implies infinite loop gain. Nonlinear analysis as a bang-bang, sampled loop is thereby imposed upon us. The nonlinear loop can be very useful despite the analytical difficulties.)

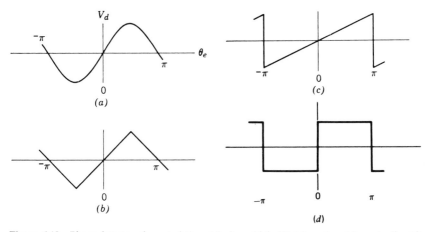

Figure 6.12 Phase-detector characteristics: (*a*) sinusoidal; (*b*) triangular. (*c*) sawtooth; (*d*) rectangular.

A sawtooth characteristic results from sampling a sawtooth input waveform, but that characteristic is more readily obtained from sequential PDs, as is explained in the next section.

Sketches of several simple PD characteristics are shown in Figure 6.12.

There are PD circuits that produce nonsinusoidal PD characteristics even when fed with sinusoidal inputs. The best known is *Tanlock*[21,22] which is implemented as in Figure 6.13. Its PD characteristic is given by

$$V_d = K_d \frac{(1+y)\sin\theta_e}{1+y\cos\theta_e} \tag{6.15}$$

which is plotted for several values of y in Figure 6.14a. The dimensionless factor y is selected between 0 and 1. If $y=0$ we have the ordinary sinusoidal PD. If $y \geqslant 1$ the loop becomes unstable (actually, the division fails).

Tanlock interconnects a pair of conventional phase detectors to produce an extended PD characteristic—one that has a wider linear region and permits better hold-in and tracking performance.

The analog divider in a Tanlock circuit tends to be an awkward item; it is mechanized by placing an analog multiplier in the feedback path of an operational amplifier. Dynamic range and stability problems must be overcome.

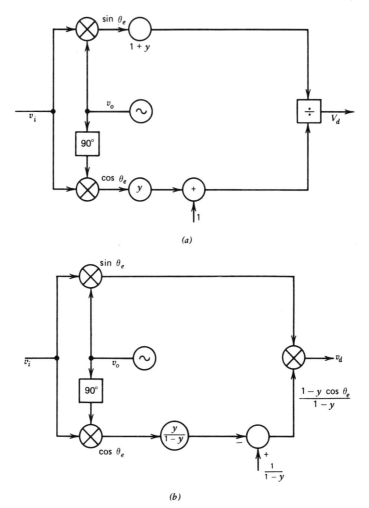

Figure 6.13 Modified phase detectors: (a) Tanlock phase detector; (b) detector-product phase detector.

The nuisance of a divider can be avoided by a circuit that generates

$$V_d = K_d \sin \theta_e \frac{(1 - y \cos \theta_e)}{1 - y} \tag{6.16}$$

by means of an extra (cosine) phase detector and an analog multiplier. As shown in Figure 6.14b, its linearity extension is not as good as that of Tanlock.

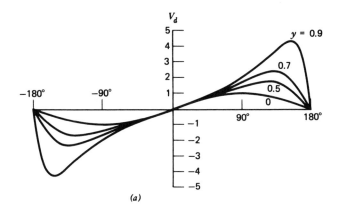

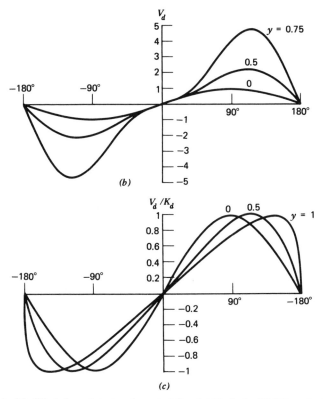

Figure 6.14 Modified phase-detector characteristics: (a) Tanlock; (b) PD product; (c) phase feedback.

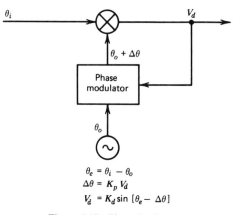

$$\theta_e = \theta_i - \theta_o$$
$$\Delta\theta = K_p V_d$$
$$V_d = K_d \sin\left[\theta_e - \Delta\theta\right]$$

Figure 6.15 Phase feedback.

A third circuit[23] (Figure 6.15) feeds back the PD output voltage into a phase modulator, thereby reducing the phase error between the two signals actually applied to the phase detector. Its range extension is shown in Figure 6.14c. Remarkably, feeding the PD output voltage through an ideal differentiator into the VCO control terminal has exactly the same effect as phase feedback.[24]

These complex nonsinusoidal circuits are rarely used. One reason is that a sawtooth characteristic is readily obtained with very simple sequential circuits, as described in the next section. A sawtooth is superior in linearity and range extension to any of the illustrated characteristics.

Another reason, which is explored later in the chapter, is that noise degenerates any extended characteristic. If the signal is immersed in noise, the PD characteristic approaches sinusoidal irrespective of its shape for signal alone.

A sinusoidal PD characteristic has the same magnitude of slope at the unstable null at 180° as it does at the stable null at 0° (see Figure 5.1). The same is true for a triangular or rectangular characteristic or any PD with even symmetry about its peak output. Feedback polarity with this type of PD is immaterial; the loop selects automatically whichever of the two nulls provides negative feedback.

An extended PD characteristic (sawtooth, Tanlock, etc.) has unequal magnitudes of slope at the two nulls. To assure stable tracking about the desired null, the polarity around the entire loop must be correct. Reverse polarity forces the loop to try to track about the wrong null.

Sequential Phase Detectors

This important class of circuits operates on the zero crossings of the signal and local oscillator; any other characteristics of the waveforms are ignored. For reliable operation of the circuits, the waveforms are usually clipped to a rectangular shape.

Average output is proportional to the time interval between a level transition of the signal and a transition of the VCO waveform. The circuit must have some memory to measure the time difference.

The simplest sequential PD is an ordinary RS flip flop. Negative transitions on one input set the flip flop to a true state and negative transitions on the other input reset it to the false state. Typical waveforms are shown in Figure 6.16 and the PD characteristic—a sawtooth—is in Figure 6.17.

We denote by θ_d the phase difference between input signal and VCO output. Useful output is the DC average on one output terminal of the flip flop and is denoted by V_d. For $0 < \theta_d < 2\pi$,

$$V_d = V_H \theta_d \qquad (6.17)$$

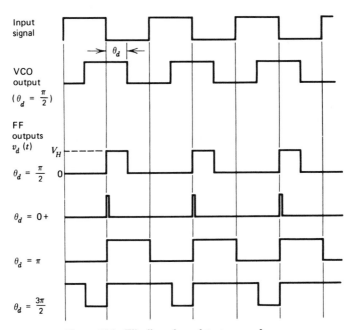

Figure 6.16   Flip-flop phase-detector waveforms.

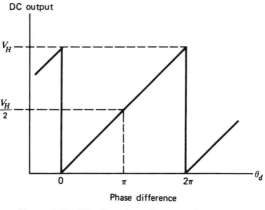

Figure 6.17 Flip-flop phase-detector characteristic.

where V_H is defined in Figure 6.16. The linear range is centered at $\theta_d = 180°$, rather than at 90° as found in multiplier PDs. Equilibrium tracking is ordinarily centered around 180°, so the DC offset of $V_H/2$ must be cancelled out with an appropriate bias circuit.

(Digital integrated circuits are manufactured without regard to small noise voltages that might appear on the high or low logic levels of individual devices. If low circuit noise—high spectral purity—is needed, then it is advisable to use the digital circuit to drive a low-noise, discrete-component analog gate to produce the actual DC output. Offset bias can be compensated in the auxiliary gate.)

Ripple is clearly a square wave at the signal frequency and has a duty ratio that depends on phase error.

The flip flop need not be operated at the actual input frequency; digital counters can divide the input frequency by a factor N.[25] The linear phase range of the PD, referred to the input phase, becomes $2\pi N$, which is in strong contrast to the much smaller range achievable with a sinusoidal, multiplier PD.

Let us suppose that the input signal fails. Then the next VCO negative transition will reset the flip flop and it stays reset until the signal returns. The loop interprets the steady reset condition as a large phase error and attempts to correct it by lowering the VCO frequency. Eventually, the loop filter or VCO is pushed against a saturation limit and remains in this condition.

Signal failure illustrates a general problem of sequential phase detectors; the circuit tends to be intolerant of missing or extra transitions. This behavior should be contrasted to that of multipliers in which the transition, as such, has little influence; the total integrated waveform determines the DC output. We see later how this transition-sensitive property affects PD operation in the presence of noise.

The problem caused by input signal failure is easily remedied in the simple flip flop: simply arrange the circuit so that the VCO transitions toggle the flip flop rather than reset it.[25] Then if the input fails, the flip flop toggles back and forth between the two logic levels with a 50% duty ratio, which the loop interprets as zero phase error. The loop then tends to remember its existing state and is prepared to quickly reacquire the signal upon its return.

A widely used sequential PD of greater complexity is shown in Figure 6.18. It consists of four flip flops plus additional logic and is available in several versions as a single-chip integrated circuit.[26,27] It is called a *phase-frequency detector* (PFD) because it also provides an indication of frequency error when the loop is out of lock.

The PFD has two output terminals, labeled U and D. The low (pulled-down) condition is active; the high condition is inactive on each terminal.

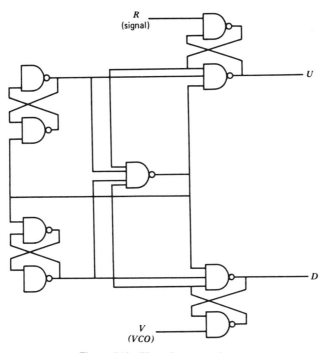

Figure 6.18 Phase-frequency detector.

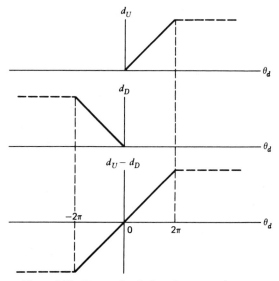

Figure 6.19 Duty ratio of phase-frequency detector.

Both U and D can be high simultaneously, but not low. Duty ratio, d_U or d_D, is the fractional time either terminal is in the low condition.

We denote the phase difference between signal and VCO as θ_d; the duty ratios as a function of θ_d are sketched in Figure 6.19. Several unique properties can be seen:

- Active phase range is $\pm 360°$, which is double that of other PDs. It is linear over the entire range.
- The PD characteristic is *aperiodic*, whereas all others have been periodic in θ_d.
- If the loop is unlocked, only one output terminal—U or D, as appropriate—pulls down into the active condition. The low terminal indicates the direction of frequency error and thereby automatically provides sweep-frequency acquisition capability.
- Both outputs are quiescent at the equilibrium tracking point $\theta_d = 0$. In the near vicinity of equilibrium, one or the other output pulls down with small duty cycle; the ripple is only a narrow pulse at the input frequency. Narrow pulses are much easier to filter than square waves.
- There may be some crossover distortion in the PD characteristic around $\theta_d = 0$.

Operation of the circuit can be analyzed by examining the 12 different logic states possible and the transitions between the states caused by the two inputs. Operation in a locked condition can be followed exhaustively by such an analysis.

In the unlocked condition, the analysis indicates that one terminal (U or D, depending on the sign of the frequency difference) will remain high as long as the loop is out of lock and the other terminal switches back and forth between levels in a manner determined by the frequency and phase differences. One might consider that a sort of chopped beat-note appears at the active terminal.

General characteristics of the beat-note are obscure; sequential state analysis provides the fine details, but a broader overview remains hidden in the details. A duty ratio of 50%, on the average, might be expected intuitively. One fact is certain: the active terminal does not simply clamp low as might be inferred from casual reading of the manufacturers' literature.

Note that the detector provides an average slew signal to the loop filter when out of lock. The acquisition aid provided therefore is akin to frequency sweeping and not discriminator aiding nor improved pull-in (see Chapter 5).

If a signal transition is missing, or if an extra one appears, the PFD interprets this event as a loss of lock and tries to reacquire lock. Since it has its own memory, the effects of an extra (missing) transition propagate for more than one cycle. If the loop is tracking with small error, a missing transition will cause a very large error voltage to appear for a short time. It is clear that the PFD is intolerant of missing or extra transitions; the implications for operation in noise are considered presently.

6.4 LIMITERS

An introduction to limiters is needed before we examine the effects of noise on phase-detector operation. We restrict our attention to an ideal, *bandpass, hard* limiter. It is bandpass because a narrowband filter, centered at the signal frequency, precedes it. A hard limiter has an input voltage v_i and output

$$v_L = V_L \operatorname{sgn}(v_i) \tag{6.18}$$

which is a rectangular waveform that preserves the locations of the zero crossings of the filtered input. A *zonal filter* may follow the limiter to remove all harmonics and pass only the fundamental band.

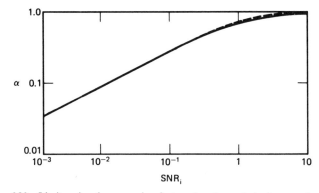

Figure 6.20   Limiter signal-suppression factor: (---) exact; (—) approximation.

Limiter action can be analyzed for an input consisting of a sinusoidal signal plus gaussian noise.[28-30] Various interesting properties are revealed by the analysis.

Output power from the limiter is constant, irrespective of input signal-to-noise ratio. Since the output waveform is a square wave of constant amplitude, this result is hardly surprising; the only effect of noise is to cause jitter of the zero crossings of the square wave. However, the output power in each zone (that is, harmonic band—fundamental, third harmonic,* fifth harmonic, etc.) is constant irrespective of input SNR.

In the absence of noise, the fundamental output is a sine wave with amplitude $4V_L/\pi$. When noise is added to the input, the output signal level must decrease because the total output signal plus noise is held constant; noise suppresses the signal in a limiter. Signal suppression is given the symbol α and is a function of the input signal-to-noise power ratio ρ_i, as measured in the passband of the input filter. We interpret α as the ratio of the fundamental signal amplitude at finite ρ_i to the amplitude $4V_L/\pi$, which obtains in the absence of noise. Signal suppression is given by

$$\alpha = \sqrt{\frac{\pi \rho_i}{4}} \left[I_o\!\left(\frac{\rho_i}{2}\right) + I_1\!\left(\frac{\rho_i}{2}\right) \right] e^{-\rho_i/2}$$

$$\approx \sqrt{\frac{\rho_i}{\rho_i + 4/\pi}} \tag{6.19}$$

---

*A symmetrical limiter does not produce even harmonics.

where I_0 and I_1 are modified Bessel functions; the ratio α is plotted in Figure 6.20. The approximate formula is more than accurate enough for engineering calculations.

The gain K_d of a multiplier-type phase detector is proportional to the applied signal voltage. If the signal voltage is suppressed by a factor α, then K_d is reduced similarly. Therefore, loop gain and bandwidth are a function of input signal-to-noise ratio if a limiter precedes the phase

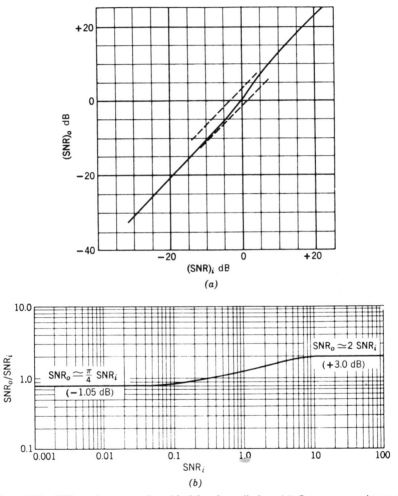

Figure 6.21 SNR performance of an ideal bandpass limiter. (*a*) Output versus input. (*b*) Noise degradation. (SNR$_i$ = ρ_i.)

detector. Suppression is a major effect of a limiter and must be taken into account in the loop calculations.

Output signal-to-noise ratio SNR_o in the fundamental zone is also very interesting. Analyses show that the output SNR_o is degraded by no more than 1.05 dB at very low input SNRs and shows an improvement of 3 dB at very large input SNRs! Its behavior is plotted in Figure 6.21.

These SNR_o results are correct but cannot be applied uncritically to the analysis of PLLs, contrary to early thinking. The 3-dB improvement was the first item to be recognized as unrealistic; even though a limiter does indeed improve SNR_o by 3 dB for large input SNRs, that improvement does not accrue in any way as reduced phase jitter in the PLL.

High-SNR improvement reflects the suppression of the AM component of noise; the limiter has no influence on the PM component. Since PLL jitter depends only on phase, not amplitude, the suppression of AM noise does not improve loop tracking.

Moreover, jitter degradation at low input SNR is not as bad as 1.05 dB. The limiter spreads the noise spectrum of the input[30] so that the output spectrum of the fundamental zone has relatively increased density in the tails and decreased density at the center of the spectrum. A narrowband PLL passes mainly the central portion of the spectrum, so noise degradation is less than 1.05 dB. The true degradation depends on the input filter shape, the post filter, and the PD configuration. Further discussion is deferred to the next section.

6.5 NOISE EFFECTS ON PHASE DETECTORS

Noise has many adverse effects on operation of phase detectors. One arises from the unavoidable DC offsets in the loop, particularly in the PD itself. Offsets arise from uncompensated biases, unbalanced circuits, rectified noise, incidental frequency discrimination, and a host of even more esoteric sources. Offset is usually dependent on temperature, frequency, SNR, and time.

If a limiter is used, the signal amplitude is suppressed at low input SNR. If limiting is not used, the signal amplitude must be small so that signal plus worst-case noise does not overload the PD. In either case, the noise-caused offset increases with worsening SNR.

If the useful output of the PD is so small that it cannot overcome the offset, tracking fails and the loop loses lock. This occurrence is dubbed *phase-detector threshold* and is caused by unavoidable defects in the circuits rather than any inherent property of a PLL. Nonetheless, any real phase-detector circuit has such defects and they must be taken into account in the design.

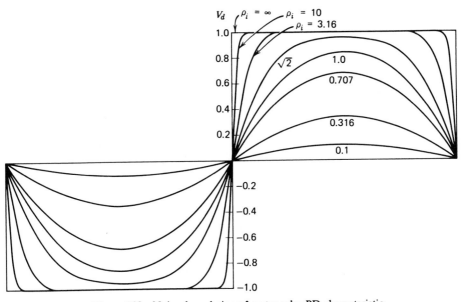

Figure 6.22 Noise degradation of rectangular PD characteristic.

A poorly balanced phase detector might exhibit PD threshold for input SNR of about -20 dB, while a well-designed circuit might tolerate -30 dB. Painstaking design efforts are needed to obtain satisfactory operation below about -25 dB. Methods of controlling input SNR are given in Chapter 8.

Another effect is the degeneration of a nonsinusoidal PD characteristic to a sinusoidal characteristic in the presence of large input noise. Pouzet [31] has proven that any periodic PD* characteristic loses its noisefree shape and tends towards sinusoidal as the input SNR becomes small. Figure 6.22 shows an example for a rectangular characteristic, but similar changes [31,32] occur for any of the other common shapes.

Shape of a PD characteristic at arbitrary SNR can be calculated by Pouzet's analysis. Physical insight is gained by realizing that the phase of signal plus noise fluctuates randomly about the mean phase, which is that of the signal alone. Useful DC output may be regarded as the fluctuating input phase averaged over the nonlinear noisefree PD characteristic.

We represent the mean phase error by θ_e and the noisefree PD characteristic by $C(\theta_e)$. The phase fluctuation caused by the noise is designated θ and has probability density $p(\theta)$, a function of ρ_i. The resultant phase of

*The sequential phase-frequency detector has an aperiodic characteristic; its degeneration in large noise is unknown.

signal plus noise is $\theta_e - \theta$. Average DC output of the phase detector is

$$V_d(\theta_e, \rho_i) = \int_{-\pi}^{\pi} C(\theta_e - \theta) p(\theta) \, d\theta \qquad (6.20)$$

where θ is taken modulo 2π.

Expressed in this manner, the DC output V_d is seen to be the convolution of the noisefree characteristic $C(\theta_e)$ by the input phase probability density $p(\theta)$. In the absence of noise, the phase density is a delta function $\delta(\theta)$ and the DC output reduces to $V_d(\theta_e, \infty)$. When noise is present, the convolution causes the DC output to be a smeared and diminished version of $C(\theta_e)$.

If $p(\theta) = p(-\theta)$, as is true if the input noise is gaussian, and if $C(\theta_e) = -C(-\theta_e)$, then the null at $\theta_e = 0$ will not shift with varying input SNR. If $C(\theta_e)$ is not odd-symmetric, then the null can shift—a highly unsatisfactory occurrence.

Not only does noise cause the PD characteristic to degenerate towards sinusoidal, but the slope at the null is reduced; this is signal suppression and is described explicitly in Section 6.4 for a sinusoidal PD preceded by a limiter. To find suppression for other characteristics, one differentiates (6.20) with respect to θ_e and evaluates the integral at $\theta_e = 0$. From Pouzet's paper one can infer that suppression in any ordinary PD will not deviate radically from that found for the sinusoidal PD.

The various piecewise-linear characteristics all require that a limiter precede the phase detector. A triangular characteristic is obtained if the square-wave limiter output is unfiltered and is used to drive a switching phase detector. A sawtooth characteristic is obtained if the unfiltered limiter output drives a suitable sequential PD, while a rectangular characteristic results if the unfiltered limiter output is sampled synchronously.

The characteristic is sinusoidal if either input to a PD is sinusoidal. This can occur if the input bandpass signal is not limited, if the limiter output is filtered to remove harmonics, or if the VCO drive to the PD is sinusoidal. All three alternatives yield identical performance.

We state earlier that a limiter degrades signal-to-noise ratio for small ρ_i. Phase jitter of the PLL must also be worsened, but by an amount other than the total SNR loss. Jitter increase depends on the shapes of the PD characteristic and the input bandpass filter. Pouzet has calculated the increase for various conditions; his results are summarized in Table 6.1. The numbers shown represent the asymptotic increase of jitter for very low input SNR.

Very little loss is incurred with a sinusoidal PD or even a triangular or rectangular PD, especially if the input filter is single tuned. (It becomes

Table 6.1   PLL Jitter Increase Due to Limiter (For $\rho_i \ll 1$)

PD Type	Bandpass Filter	
	Single-Tuned (dB)	Rectangular (dB)
Sinusoidal (No limiter)	0	0
Sinusoidal (With limiter)	0.25	0.65
Triangular	0.3	0.7
Rectangular	0.36	0.97
Sawtooth	2.9	2.9

apparent in Chapter 8 that single-tuned filters are preferable for other reasons as well.) However, there is a severe loss with a sawtooth characteristic. Since the actual characteristic degenerates to sinusoidal anyhow, it is difficult to justify the use of a sawtooth PD if the input signal is normally immersed in the noise. Similar results ought to be anticipated from any other extended PD characteristic.

As a final topic, we reconsider sequential phase detectors. It is explained earlier that these devices, and most particularly the phase-frequency detector, can be intolerant of extra or missing zero crossings. Large noise can induce extra or missing crossings; in general, the number of crossings depends on the noise spectrum and the signal-to-noise ratio.[33] If the number of crossings per second departs from the signal frequency, the phase-frequency detector acts as though the loop is out of lock and tries to slew the VCO frequency to bring the loop back to "lock." At the very least, the wrong number of crossings will cause an additional bias of the PD output and, if the number of crossings is sufficiently wrong, tracking will fail entirely. A sequential PD should be used in a noisy environment only with great caution and for well-justified reasons.

This problem is mitigated if the noise spectrum is shaped so that the rate of noise crossings is equal to the signal frequency. A spectrum with arithmetic symmetry about the signal frequency has the desired property.

6.6 DIGITAL IMPLEMENTATIONS

Despite the occasional appearance of digital circuits, particularly in sequential phase detectors, all the foregoing material has dealt with analog PLLs. To be classified as *digital*, a loop ought to have at least these two properties:

1. Output phase is generated in discrete increments, not as a continuous function.
2. Error signal is generated as a digital number, not as an analog voltage.

Many loops obeying these criteria have been built; a sampling of the literature may be found in Refs. 34–40.

Digital PLLs come in a great variety of forms and can be applied where analog loops are difficult or impossible. Methods of analysis differ widely from the analog methods presented in this book. When the subject matures, another book of this size could probably be written on digital implementations alone.

Rather than attempt to explain digital loops in any detail, we restrict the presentation here to a couple of brief examples. Much of the underlying analysis of digital loops remains to be performed.

Figure 6.23 is a block diagram of a rather complex digital PLL; the actual digital operations might be performed by a computer. Operation is as follows:

- Input is sampled at a fixed sampling rate, not generally synchronous with the signal frequency. To avoid aliasing[41] at this point, the sampling rate should exceed double the bandwidth of the input.
- Samples are quantized and converted into digital numbers in the analog-to-digital convertor (ADC).

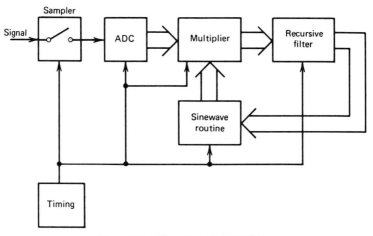

Figure 6.23 Complicated digital PLL.

- Digital signal samples are mutiplied by digital samples of a sine wave of computed phase. Care must be taken to avoid trouble from aliasing of ripple. Multiplier output is equivalent to PD output from an analog PLL.
- A digital filter takes the place of the analog filters discussed heretofore and also the integrator associated with the analog VCO.
- Output of the digital filter is the phase of the digital "VCO." Note that the familiar analog components tend to merge with one another in a digital implementation.
- Output of the "VCO" is computed in a sine wave routine and is used to multiply the digitized input samples.

The foregoing example is rather complicated and cannot be implemented casually. It has the advantage that the incoming signal can be recorded in digital form, on tape for example, and then processed later at a more convenient time and place. Imperfections of tape recorders (primarily flutter) often obstruct recording and playback of analog signals.

A second example, by contrast, is extremely simple, as seen in Figure 6.24. It consists only of a sampler, a counter, and a fixed-rate clock oscillator. The sampler operates virtually synchronously with the input signal and merely samples the input polarity. From previous discussion, we recognize that the PD characteristic is rectangular and the loop works in a bang-bang mode. Again, sampling rate must be sufficient to avoid aliasing.

Input frequency is f_i, oscillator frequency is f_x, and the counter counts down the oscillator frequency to f_x/N, in the absence of loop feedback. The loop is designed for $f_x/N \cong f_i$.

The counter is arranged so that its count is advanced or "bumped" by one increment if the sample is positive and retarded by one increment if

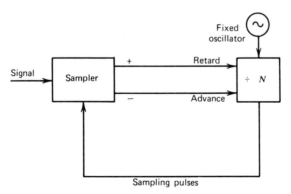

Figure 6.24 Simple digital PLL.

the sample is negative. Each time the counter runs down to zero, it generates a sampling pulse. Phase at the counter output changes in increments of $1/N$ cycle. (Bump increments within the counter can be smaller.[40]) Output phase has been quantized, thereby creating a source of phase jitter not present in an analog loop.

The simplified digital loop can be shown to be similar to a nonlinear, first-order, analog PLL. The illustrated method of shifting output phase is very popular among digital PLLs and is variously called a digital phase shifter or an incremental phase shifter. Simple digital PLLs of this sort are well suited for realization as special-purpose digital hardware.

REFERENCES

1. E. A. Gerber and R. A. Sykes, "Quartz Crystal Units and Oscillators", *Time and Frequency: Theory and Fundamentals*, NBS Monograph 140, U.S. Department of Commerce, May 1974, Chap. 2.

2. E. P. Felch and J. O. Israel, "A Simple Circuit for Frequency Standards Employing Overtone Crystals," *Proc. IRE*, Vol. 43, pp. 596–603, May 1955.

3. W. L. Smith, "Miniature Transistorized Crystal-Controlled Precision Oscillators," *IRE Trans.*, *I-9*, pp. 141–148, September 1960.

4. M. M. Driscoll, "Two-Stage Self-Limiting Series Mode Type Quartz-Crystal Oscillator Exhibiting Improved Short-Term Frequency Stability," *IEEE Trans.*, *IM-22*, pp. 130–138, June 1973.

5. J. A. Barnes, *Models for the Interpretation of Frequency Stability Measurements*, NBS Technical Note 683, U.S. Department of Commerce, August 1976.

6. D. W. Allan, J. H. Shoaf, and D. Halford, "Statistics of Time and Frequency Analysis" *Time and Frequency: Theory and Fundamentals*, NBS Monograph 140, U.S. Dept. of Commerce, May 1974, Chap. 8.

7. D. Halford, A. Wainwright, and J. Barnes, "Flicker Noise of Phase in R. F. Amplifiers and Frequency Multipliers: Characterization, Cause, and Cure," *Proc. 22nd Symp. Freq. Control*, 1968, pp. 340–341.

8. W. R. Attkinson, L. Fey, and J. Newman, "Spectrum Analysis of Extremely Low Frequency Variations of Quartz Oscillators," *Proc. IRE*, Vol. 51, p. 379, February 1963.

9. R. M. Gray and R. C. Tausworthe, "Frequency-Counted Measurements and Phase Locking to Noisy Oscillators," *IEEE Trans.*, *COM-19*, pp. 21–30, February 1971.

10. W. C. Lindsey and C. M. Chie, "Theory of Oscillator Instability Based Upon Structure Functions," *Proc. IEEE*, Vol.64, pp. 1652–1665, Dec. 1976.

11. W. C. Lindsey and C. M. Chie, "Theory of Oscillator Phase Noise in Correlative Tracking Systems", *Conference Record of the International Communications Conference*, Vol.3, 1978, pp. 50.5:1–50.5:8.

12. F. L. Walls and A. E. Wainwright, "Measurement of the Short-Term Stability of Quartz Crystal Resonators and the Implications for Crystal Oscillator Design and Applications," *IEEE Trans.*, *IM-24*, pp. 15–20, March 1975.

13. P. Grivet and A. Blaquiere, "Non-Linear Effects of Noise in Electronic Clocks," *Proc. IEEE.*, Vol. 51, pp. 1606–1614, November 1963.

14. B. Gilbert, "A Precise Four-Quadrant Multiplier with Subnanosecond Response," *IEEE J.*, SC-3, pp. 365–373, December 1968.

15. A. Blanchard, *Phase-Locked Loops*, Wiley, New York, 1976, Chap. 2.

16. W. J. Gruen, "Theory of AFC Synchronization," *Proc. IRE*, Vol. 41, pp. 1043–1048, August 1953.

17 F. M. Gardner, *Phaselock Techniques*, 1st ed. Wiley, New York, 1966, Chap. 5.

18 C-S. Yen, "Phase-Locked Sampling Instruments," *IEEE Trans.*, *IM-14*, pp. 64–68, March–June 1965.

19. B. R. Eisenberg, "Gated Phase-Locked Loop Study," *IEEE Trans.*, *AES-7*, pp. 469–477, May 1971.

20. S. Barab and A. L. McBride, "Uniform Sampling Analysis of a Hybrid Phase-Locked Loop with a Sample-and-Hold Phase Dectector," *IEEE Trans.*, *AES-11*, pp. 210–216, March 1975.

21. L. M. Robinson, "TANLOCK: A Phase-Lock Loop of Extended Tracking Capability," *Proc. IRE Conv. Mil. Electron.*, Los Angeles, February 1962, pp. 396–421.

22. M. Balodis, "Laboratory Comparision of TANLOCK and Phaselock Receivers," *Conference Record National Telemetry Conference 1964*, Paper 5-4.

23. A. Acampora and A. Newton, "Use of Phase Subtraction to Extend the Range of a Phase-Locked Demodulator," *RCA Rev.* Vol. 27, pp. 577–599, December 1966.

24. J. Klapper and J. T. Frankle, *Phase-Locked and Frequency Feedback Systems*, Academic Press, New York; 1972, Chapter 8.

25. C. J. Byrne, "Properties and Design of the Phase-Controlled Oscillator with a Sawtooth Comparator," *BSTJ*, 41, pp. 559–602, March 1962.

26. *Phase-Locked Loop Data Book*, 2nd ed., Motorola, Inc., August 1973.

27. D. K. Morgan and G. Steudel, *The RCA COS/MOS Phase-Locked-Loop*, Application Note ICAN-6101, RCA, Somerville, NJ, October 1972.

28. W. B. Davenport, Jr., "Signal-to-Noise Ratios in Band-Pass Limiters," *J. Appl. Phys.*, Vol. 24, pp. 720–727, June 1953.

29. J. C. Springett and M. K. Simon, "An Analysis of the Phase Coherent-Incoherent Output of the Bandpass Limiter," *IEEE Trans.*, *COM-19*, pp. 42–49, February 1971.

30 J. H. van Vleck and D. Middleton, "The Spectrum of Clipped Noise," *Proc. IEEE*, Vol. 54, pp. 2–19, January 1966.

31 A. H. Pouzet, "Characteristics of Phase Detectors in Presence of Noise," *Proc. Int. Telem. Conf.*, Los Angeles, Vol 8, pp. 818–828, 1972.

32. A. Blanchard, *Phase-Locked Loops*, Wiley, New York, 1976, Chap. 7.

33. S. O. Rice, "Mathematical Analysis of Random Noise," *BSTJ*, 23, pp. 282–332, 1944; 24 pp. 46–156, 1945. Also in N. Wax (Ed.), *Noise and Stochastic Processes*, Dover, New York, 1954.

34. J. R. Cessna and M. D. Levy, "Phase Noise and Transient Times for a Binary Quantized Digital Phase Locked Loop in White Gaussian Noise," *Trans. IEEE, COM-20*, pp. 94–103, April 1972.

35. J. Garodnick, J. Greco, and D. L. Schilling, "Response of an All-Digital Phase-Locked Loop," *Trans. IEEE, COM-22*, pp. 751–763 June 1974.

36. G. S. Gill and S. C. Gupta, "First-Order Discrete Phase-Locked Loop with Applications to Demodulation of Angle-Modulated Carrier," *Trans. IEEE, COM-20*, pp. 454–462, June 1972.

37. J. K. Holmes, "Performance of a First-Order Transition Sampling Digital Phase-Locked Loop Using Random-Walk Models," *Trans. IEEE, COM-20*, pp. 119–131, April 1972.

38. J. K. Holmes and C. R. Tegnelia, "A Second-Order All-Digital Phase-Locked Loop," *Trans. IEEE, COM-22*, pp. 62–68, January 1974.

39. G. Pasternack and R. L. Whalin, "Analysis and Synthesis of a Digital Phase Locked Loop for FM Demodulation" *BSTJ*, 47, pp. 2207–2239, December 1968.

40 J. R. Cessna, "Digital Phase Locked Loops with Sequential Filters," *Proc. Int. Telem. Conf., Los Angeles*, Vol. 8, pp. 136–148, 1972.

41 L. R. Rabiner and B. Gold, *Digital Signal Processing*, Prentice-Hall, Englewood Cliffs, NJ, 1975, Chap 2.

Chapter Seven

Optimization of Loop Performance

Two general principles may be abstracted from the preceding chapters:

1. To minimize output phase jitter due to external noise the loop bandwidth should be made as narrow as possible.
2. To minimize transient error due to signal modulation, to minimize output jitter due to internal oscillator noise, or to obtain best tracking and acquisition properties, the loop bandwidth should be made as wide as possible.

These principles are directly opposed to one another; improvement in one type of performance can come only at the expense of degrading the other, and therefore some compromise between the two is always necessary. Almost always there is a compromise that is "best" in some sense; this compromise is called "optimum."

It must be recognized that there is no unique optimum result that applies under all conditions. On the contrary, there are many possible results, depending on the criteria of performance, the nature of the input signal, and any restrictions placed on loop configuration.

The best-known optimization is that derived by Jaffe and Rechtin[1] following the Wiener method.* Their criterion of loop performance is the mean-square loop error

$$\Sigma^2 = \overline{\theta_{no}^2} + \lambda^2 E_T^2 \tag{7.1}$$

where $\overline{\theta_{no}^2}$ is the phase jitter due to noise (3.18), and E_T^2 is a measure of the

---

*Details of the Wiener method are far beyond the scope of this book. For an extensive exposition of the subject, see Y. W. Lee, *Statistical Theory of Communication*, Wiley, New York, 1960, Chaps. 14–17.

total transient error:

$$E_T^2 = \int_0^\infty \theta_e^2(t)\,dt \qquad (7.2)$$

where $\theta_e(t)$ is the instantaneous phase error in the loop due to transients. The quantity λ is a Lagrangian multiplier that establishes the relative proportions of noise and transient error that are to be permitted. [Notice that λ^2 has dimensions of (time)$^{-1}$—that is, frequency.]

In the Wiener optimization method the known quantities are the spectra of the signal and noise, whereas the criterion of performance is the mean-square error Σ^2. The result of the method is a description of an "optimum" filter whose output provides a minimum mean-square error.

Jaffe and Rechtin have assumed white noise and three different types of modulation at the input: phase step, frequency step, and frequency ramp. For each condition they arrive at an optimum loop transfer function $H(s)$ and the corresponding transfer function for the loop filter $F(s)$. Results are summarized in Table 7.1.

For the three different types of input the optimum filter types are first-, second-, and third-order loops, respectively; the Wiener method specifies optimum filter shape as well as bandwidth. In the optimum second-order loop (of greatest interest because of its widespread usage) damping factor is $\zeta = 0.707$. The optimum third-order loop has complex zeros, which are not usually convenient to mechanize nor well-suited for reliable acquisition.

Note that optimum bandwidth is a function of the input signal-to-noise ratio. To minimize the total error the loop should be capable of measuring

Table 7.1 Wiener—Optimized Loops

Input	Optimum $H(s)$	$F(s)$	Optimum Bandwidth*
Phase Step			
$\theta_i(t) = \Delta\theta$	$\dfrac{\omega_1}{s + \omega_1}$	$\dfrac{\omega_1}{K_o K_d}$	$\omega_1 = \Delta\theta\lambda\left(\dfrac{2P_s}{W_o}\right)^{1/2}$
Frequency step			
$\theta_i(t) = \Delta\omega t$	$\dfrac{\omega_n^2 + \sqrt{2}\,\omega_n s}{\omega_n^2 + \sqrt{2}\,\omega_n s + s^2}$	$\dfrac{\omega_n^2 + \sqrt{2}\,\omega_n s}{K_o K_d s}$	$\omega_n^2 = \Delta\omega\lambda\left(\dfrac{2P_s}{W_o}\right)^{1/2}$
Frequency ramp			
$\theta_i(t) = \dfrac{\Delta\dot{\omega} t^2}{2}$	$\dfrac{\omega_3^3 + 2\omega_3^2 s + 2\omega_3 s^2}{\omega_3^3 + 2\omega_3^2 s + 2\omega_3 s^2 + s^3}$	$\dfrac{\omega_3^3 + 2\omega_3^2 s + 2\omega_3 s^2}{K_o K_d s^2}$	$\omega_3^3 = \Delta\dot{\omega}\lambda\left(\dfrac{2P_s}{W_o}\right)^{1/2}$

*P_s = input signal power; W_o = input noise spectral density (one-sided).

SNR and readjusting its bandwidth for optimum performance. To perform this optimum adaptation exactly would be a complex and difficult task; as far as is known, there has never been a serious attempt at perfect adaptation.

One reason for the lack of effort is that Jaffe and Rechtin discovered that near-optimum adaptation may be achieved by very simple means, namely, use of a bandpass limiter before the phase detector. In Chapter 6, it is shown that the presence of a limiter causes loop bandwidth and damping to vary as a function of input SNR. This variation is not optimum (damping should remain constant, and the variation of ω_n should have a different form), but it is sufficiently close to optimum to be useful. Limiters are widely used in sensitive phaselock receivers.*

It is of interest to observe that the definition of E_T^2 given here is such that steady-state error must be zero. If this were not true, E_T^2 would be infinite. If some other definition of transient error were to be used (e.g., peak error), it is probable that different optimum results would be obtained.

The Wiener analysis is strictly applicable only to linear systems; to apply it to the phaselock loop requires that the linear approximation be made. Furthermore, Jaffe and Rechtin's exact results are applicable only if noise is white, when the input is one of the three specific types listed here, and when the error criterion is that given in (7.1); all this is to say that we have so far shown only *an* optimum (or rather three optimum loops) and not *the* optimum loop, even in the restricted category of Wiener filters.

In practice, when narrow bandwidth is needed, a second-order loop is the type most commonly used. A first-order loop necessitates a major sacrifice of hold-in range and has poor phase-slope properties, whereas a third-order loop is more complicated and harder to analyze and can become unstable if not treated properly. (However, both first- and third-order loops have their uses in which they substantially outperform the second-order loop.) For the remainder of this chapter we restrict ourselves to the second-order loop and give examples of the different optimizations that are possible.

Let us suppose that the natural frequency is determined by some well-defined dynamic feature of the input signal. For example, a satellite signal exhibits a very definite rate of change of Doppler frequency; if a limit is placed on the permissible acceleration error, ω_n is immediately

---

*It should be possible to obtain similar performance from wideband (noncoherent) AGC, for the same phenomenon of signal suppression occurs. There is a slight advantage in favor of wideband AGC because the limiter causes SNR degradation and the AGC does not. Coherent AGC, on the other hand, maintains signal level constant at the phase detector and therefore has no adaptive bandwidth properties.

fixed. Given this value of ω_n, what value of damping factor results in the least phase jitter due to noise? The answer, referring to Figure 3.3, is obviously $\zeta = 0.5$, for this is the value that minimizes noise bandwidth.

For another possibility, let us suppose that noise bandwidth is fixed by, say, restrictions on the maximum allowable phase noise jitter. What value of damping will permit the largest frequency step $\Delta\omega$ without the loop being pulled out of lock, even temporarily? In Table 3.1 the noise bandwidth is shown to be

$$B_L = \frac{\omega_n}{2}\left(\zeta + \frac{1}{4\zeta}\right)$$

and (4.15) approximates pull-out frequency as

$$\Delta\omega_{PO} = 1.8\omega_n(\zeta + 1)$$

Elimination of ω_n between these equations yields

$$\Delta\omega_{PO} = \frac{3.6B_L(\zeta + 1)}{\zeta + 1/4\zeta} \tag{7.3}$$

By differentiating $\Delta\omega_{PO}$ with respect to ζ, setting the derivative equal to zero, and solving, we obtain $\zeta = 0.81$ as the damping that maximizes pull-out frequency. This maximum value is $\Delta\omega_{PO} \approx 5.82B_L$ rad/sec. Pull-out frequency is $5.79B_L$ at $\zeta = 0.707$, $5.40B_L$ at $\zeta = 0.5$, and $5.76B_L$ at $\zeta = 1.0$, so that it is hardly worthwhile to bother to optimize pull-out as such.

This finding tends to illustrate a common property of optima: the performance criterion quantity tends to change very slowly near the optimum, so that there is no need to adjust the loop to attain exactly the best performance. The extremum is usually quite broad.

These examples, plus others gleaned from the preceding chapters, are summarized in Table 7.2.

Hoffman[2] has derived another optimization that appears to be of value. A phaselock receiver is often required to track an accelerating transmitter (either true acceleration of a missile or apparent acceleration of a satellite) with a second-order loop. What acceleration error*—and, therefore, what loop bandwidth—should be used to achieve "optimum" performance?

First it is necessary to arrive at a criterion of performance. Equations 7.1 and 7.2 cannot be applied because the nonzero, steady-state acceleration

---

*If acceleration error is a serious problem, consideration should be given to a third-order loop to reduce the steady-state error to zero.

Table 7.2   Optimization of Damping in Second-Order Loops

Criterion	Constraint	Parameter ζ
Minimize noise bandwidth B_L (Table 3.1)	ω_n fixed	0.5
Minimize pull-in time T_p (5.10)	B_L fixed	0.707
Maximize sweep rate	B_L fixed	0.7–1.0
Maximize pull-out	B_L fixed	0.81
Minimize flicker jitter	B_L fixed	1.14

error would lead to an infinite integrated-square transient error. Hoffman used noise threshold as his criterion. His definition of threshold is an empirical relation taken from Martin[3] according to which at threshold

$$\theta_a + \sigma\theta_{no} = \frac{\pi}{2} \tag{7.4}$$

where θ_a is the acceleration error (4.7), θ_{no} is the rms noise jitter in the loop (3.18), and σ is a confidence factor that takes account of the fact that peak noise considerably exceeds the rms value. Equation 7.4 states that threshold error is exceeded if the sum of the individual errors exceeds 90°.

The quantity to be optimized is the input signal power P_s. From the discussion of behavior of θ_{no} in Chapter 3 and (3.20), an expression of

$$\overline{\theta_{no}^2} = \frac{\xi^2}{\mathrm{SNR}_L} \tag{7.5}$$

may be deduced. (For $\mathrm{SNR}_L > 10$, $\xi^2 = 0.5$; for $\mathrm{SNR}_L = 1$, $\xi^2 \approx 1$. The factor ξ is itself a function of SNR_L, but we regard it here as essentially constant.)

From (3.21), $\mathrm{SNR}_L = P_s/2B_L W_i$, where W_i is the input noise density. Equation 7.4 may now be written as

$$\theta_a + \sigma\xi\left(\frac{2B_L W_i}{P_s}\right)^{1/2} = \frac{\pi}{2} \tag{7.6}$$

Using Table 3.1 and (4.7), we may eliminate B_L from (7.6), leaving

$$\theta_a + \sigma\xi\left[\frac{W_i}{P_s}\left(\zeta + \frac{1}{4\zeta}\right)\sqrt{\frac{\Delta\dot{\omega}}{\theta_a}}\right]^{1/2} = \frac{\pi}{2} \tag{7.7}$$

Solving for signal power required at threshold yields

$$P_s = \frac{\sigma^2 \xi^2 (\zeta + 1/4\zeta)(\Delta\dot{\omega}/\theta_a)^{1/2} W_i}{(\pi/2 - \theta_a)^2} \tag{7.8}$$

When P_s is minimized with respect to θ_a, the surprising result is that $\theta_a = \pi/10$ (i.e., 18°), independent of σ, ξ, ζ, or W_i. This exact result depends on two approximations: that (7.4) is the definition of threshold and that ξ is constant. An exact analysis, if one should ever be discovered, would probably yield a somewhat different result, but presumably not very different.

Calculation of the minimum P_s still requires that a confidence factor σ be specified and a suitable value for ξ be deduced. The latter, which might require an iterative process, is complicated by the fact that the functional dependence of ξ on SNR_L is not known within limits closer than about ± 1 dB. Also, refer to Chapter 3 for a discussion of fundamental difficulties in defining θ_{no}.

From (7.8) it may be seen that P_s can also be minimized with respect to damping factor; the optimum value is clearly $\zeta = 0.5$. Hoffman arbitrarily uses $\zeta = 0.707$ and thereby obtains a threshold power that is higher than optimum by 0.26 dB.

Hoffman's approach suggests another possible optimization to be used when acceleration error in a second-order loop must be considered. We suppose that SNR_L is reasonably large (> 10) and let the criterion of performance be

$$\Sigma^2 = \theta_a^2 + \overline{\theta_{no}^2} = \frac{(\Delta\dot{\omega})^2}{\omega_n^4} + \frac{B_L W_i}{P_s}$$

$$= \frac{(\Delta\dot{\omega})^2 (\zeta + 1/4\zeta)^4}{16 B_L^4} + \frac{B_L W_i}{P_s} \tag{7.9}$$

which is to be minimized with respect to B_L and ζ. It is immediately evident that the optimum damping is $\zeta = 0.5$ and that the usual differentiation will yield

$$B_L^5 = \frac{P_s (\Delta\dot{\omega})^2}{4 W_i} \tag{7.10}$$

for optimum loop-noise bandwidth.

We end the chapter with one more example that may be useful. Let us suppose that the signal transmitter is essentially stationary with respect to the receiver so that dynamic phase errors may be neglected: a situation that could occur in tracking a synchronous satellite. The dominant disturbances are from additive white noise (3.18) and frequency flicker in the VCO (6.6). We seek the loop design that minimizes total phase error fluctuation

$$\Sigma^2 = \overline{\theta_{no}^2} + \overline{\theta_p^2} \tag{7.11}$$

Setting $\zeta = 1.14$ minimizes the flicker contribution and gives

$$\Sigma^2 = \frac{W_i B_L}{P_s} + 3.1 \times 10^{-2} \frac{N_1 \omega_o}{B_L^2} \tag{7.12}$$

Optimum bandwith is found to be

$$B_L^3 = 6.2 \times 10^{-2} \frac{N_1 \omega_o P_s}{W_i} \qquad \text{Hz}^3 \tag{7.13}$$

The following points summarize this chapter:

1. There is no uniquely optimum loop, nor is there a unique optimization procedure.

2. A criterion of performance must be defined. This criterion depends on the conditions of operation of the loop and the requirements placed on it. From the examples given here it may be seen that no general rule can be used in establishing the criterion.

3. Once an optimum is found, it is not usually necessary to adjust the loop parameters exactly to their optimum values. It is common for an extremum to be quite broad: in fact, to such an extent that moderate departure from optimum parameters has little adverse effect on loop performance.

REFERENCES

1. R. Jaffe and E. Rechtin, "Design and Performance of Phase-Lock Loops Capable of Near-Optimum Performance Over a Wide Range of Input Signal and Noise Levels," *IRE Trans.*, IT-1, pp. 66–76, March 1955.

2. L. A. Hoffman, *Receiver Design and the Phase-Lock Loop*, Aerospace Corp., El Segundo, CA, May 1963.

3. B. D. Martin, *The Pioneer IV Lunar Probe: A Minimum-Power FM/FM System Design*, Technical Report No. 32-215, Jet Propulsion Laboratory, Pasadena, CA, March 1962.

Chapter Eight

Phaselock Receivers
and Transponders

A simplified block diagram of a superheterodyne phaselocked receiver is shown in Figure 8.1. The incoming signal at frequency f_1 is mixed with a heterodyning local injection at frequency f_2 derived from a harmonic of the VCO at frequency f_2/N. Intermediate frequency (IF) is $f_3 = f_1 - f_2$ or $f_2 - f_1$, depending on whether low-side or high-side injection is used.

The IF amplifier may be entirely conventional, but its filtering and amplifying properties are shown in separate boxes here. A fixed oscillator at frequency f_4 is compared against IF amplifier output in a phase detector; the loop is closed through the loop filter, VCO, frequency multiplier, and mixer. In the locked condition it is necessary that $f_3 = f_4$; the input frequency therefore is $f_1 = f_2 + f_4$ or $f_2 - f_4$ for low-side or high-side injection, respectively.

Simple phaselock loops of the kind treated in earlier chapters are often known as *short loops*, while the more complex loop of Figure 8.1 is called a *long loop*, for obvious reasons.

Presence of the new elements—particularly the frequency multiplier and IF filter—can have a substantial effect on the performance and analysis of the loop. In preceding chapters gain of the VCO (K_o rad/sec-V) has been considered a property of the VCO itself, not including any subsequent multiplications. When a multiplier is used, the deviation of the VCO is increased by the multiplying factor N before injection into the mixer. Effective VCO gain is therefore really NK_o. When a multiplier is used, K_o must be replaced by NK_o wherever it appears in any equation.

8.1 EFFECTS OF IF FILTER

Inclusion of the IF filter can have much more far-reaching effects. We discuss phase-slope, noise, analysis of the filter, loop stability, and possibility of false locks.

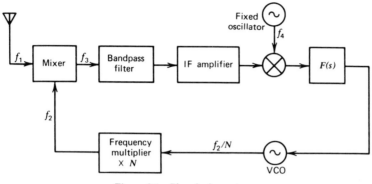

Figure 8.1   Phaselock receiver.

Phase-Slope

A narrowband IF filter necessarily has a steep phase-slope. If the frequency of the signal passing through the filter changes, the phase shift of the signal also changes. Such phase variations cause an error in the interpretation of any information being carried by signal phase; Doppler shift is an example.

In the receiver shown in Figure 8.1 the intermediate frequency in the locked-loop condition is constrained to be exactly equal to the frequency of the fixed oscillator that feeds the phase detector. For this reason, although the IF filter may itself have a steep phase-slope, the IF signal suffers no phase change as the input RF signal frequency changes.

In other mechanizations of phaselock receivers no fixed oscillator is used. Instead, a low harmonic of the VCO is injected into the phase detector, and the intermediate frequency is variable—although not nearly so much as the input frequency. Under these circumstances the signal suffers some phase change; the tolerable amount of change must be taken into account in the receiver design.

Noise

The effect of the IF amplifier on noise is best treated by means of an example. Let us suppose a signal could arise anywhere within a 500-kHz range and that the loop is designed for $B_L = 25$ Hz ($2B_L = 50$ Hz). In a simple loop, not including an IF filter, the input bandwidth to the loop would have to be at least 500 kHz merely to pass the input signal. Signal-to-noise ratio in the 500-kHz bandwidth would be 40 dB lower than SNR in the loop.

At nominal loop threshold of 0 dB the input SNR would be -40 dB. However, threshold of typical phase detectors is on the order of -30 dB, and conservative design would have them operate at no worse than -20 dB. With the wide input bandwidth, imperfections of the phase detector will limit obtainable performance, whereas properly designed equipment should be limited by inherent properties of a loop.

The typical solution to this problem is to utilize an IF bandpass filter that is much narrower than the required full input bandwidth. If, for example, an IF passband of 2.5 kHz were used, the SNR at the input to the phase detector would be only 17 dB below loop SNR, and at nominal loop threshold the phase detector would be some 13 dB above its threshold.

It should be evident that as soon as a reasonable margin against phase detector threshold has been obtained there is little to be gained by narrowing the IF bandwidth any further. Not only is a narrower filter more costly, but it can make the loop oscillate. The problem of loop stability is discussed in some detail in the following paragraphs.

IF Filter Analysis

It is first necessary to determine a method of bringing the IF bandpass filter into the loop analysis. To this end, let us consider the hypothetical test setup of Figure 8.2. We wish to determine the effect of the filter on the modulation of the test signal. Specifically, the amplitude and phase of the modulation output compared with the modulation input, as a function of modulation frequency, are desired. The result, which can be considered as a modulation transfer function denoted $F_m(s)$, is stated but not proved. If (1) the filter has a narrow, symmetrical passband, (2) the signal generator is tuned to the center frequency of the filter, and (3) the modulation deviation is very small, the approximate modulation transfer function is

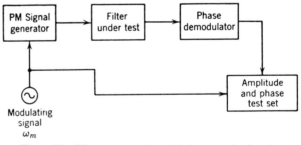

Figure 8.2 Measurement of modulation transfer function.

obtained by translating the actual filter transfer function to zero frequency and discarding the response at negative frequencies. This translation is shown in Figure 8.3.

Now let us suppose the open-loop response of the loop of Figure 8.1 were to be measured (at least conceptually, if not in actual practice) by opening the loop somewhere in its low-frequency portion and applying a low-frequency sinusoidal test signal. Total open-loop response would consist of the product of the normal response of the loop $G(s)$ and the modulation transfer function of the IF filter, or $F_m(s)G(s)$.

Stability

If the IF filter response (and therefore the modulation transfer function) is known analytically, the effect on loop response can be studied by means of a root-locus plot. Perhaps the most important filter from a practical viewpoint is a single-tuned circuit with 3-dB bandwidth of ω_B rad/sec. The equivalent modulation transfer function has a single pole at $s = -\omega_B/2$; the analysis to follow applies equally well to a short loop with an extra pole at the same location.

A root-locus plot for a second-order loop is shown in Figure 2.6. The closed-loop poles first trace out a circle for small to moderate loop gain and then remain on the negative real axis as gain increases further. One

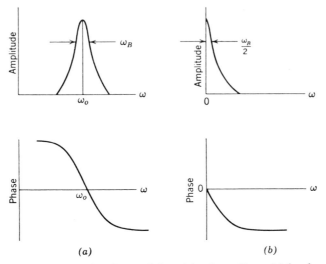

Figure 8.3 Modulation transfer characteristics of bandpass filter: (*a*) bandpass transfer function; (*b*) equivalent modulation transfer function.

root approaches infinity, while the other root approaches the open-loop zero at $s = -1/\tau_2$.

It proves to be convenient to normalize the gain as $K\tau_2$ and to define $b = \tau_2\omega_B/2$ as the normalized location of the extra pole.

The extra pole is on the negative real axis. We consider, at first, that it is far out—the IF filter has large bandwidth as compared to the PLL—so b is large. As loop gain increases, the extra pole migrates inward and eventually meets the outer PLL pole moving outward. As gain increases further, these two poles become complex with their locus asymptotic to a vertical line at $s\tau_2 = -0.5(b-1)$. (These results are obtained by standard methods of root-locus analysis that are not detailed here.)

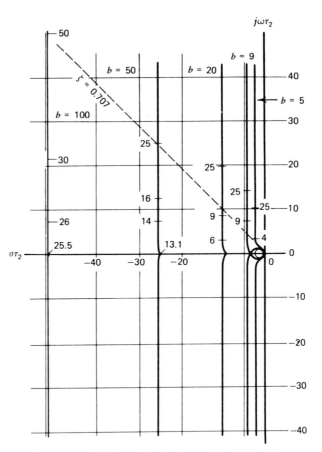

Figure 8.4 Root loci of nominal second-order loop with additional lowpass pole at $s = -b/\tau_2$. (Tick marks indicate values of $K\tau_2$.)

Representative loci are shown in Figures 8.4 and 8.5. If b is large, the extra pole does not have a great effect on the loop dynamics. Extremely large gain can lead to high-frequency ringing of the transient response, but normal values of gain produce a well-behaved loop.

If b is small, there is substantial change in performance. For b less than 9, the poles are underdamped for all values of gain; the extra pole splits open the basic circle as shown in Figure 8.5. The loop goes unstable if $b < 1$.

If a second IF pole were added to the loop, the complex asymptotes of the locus would be at $\pm 60°$ angles to the real axis. Consequently, the locus eventually enters the right half plane and the loop oscillates for large enough gain.

Behavior of a third-order loop should be essentially similar to that of a second-order loop. An extra pole at $s = -\omega_B/2$ opens up the closed locus, and the complex asymptotes are along a vertical line located at

$$s = -\left(\frac{\omega_B}{2} - \frac{1}{\tau_2} - \frac{1}{\tau_4} \right)$$

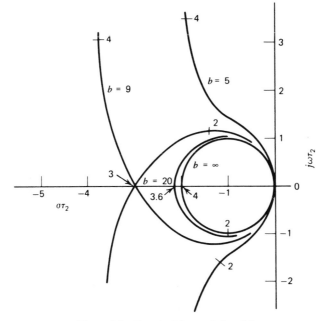

Figure 8.5   Root loci (expanded scale).

where τ_2 and τ_4 are the time constants of the lead terms of $F(s)$, assuming real zeros.

Oftentimes the response of the IF filter is not known analytically and only measured data are available, in which case a Bode plot can be used to analyze stability.

To provide an example, Figure 8.6 shows a response scaled from actual measurements on a crystal filter. The equivalent modulation transfer response is shown in Figure 8.7 along with the Bode plot of an opened second-order loop. Bandwidth of the IF filter (3 dB) is 240 rad/sec, whereas the loop has been arbitrarily chosen with $1/\tau_2 = 10$ rad/sec. Loop gain has been selected so that $\zeta = 0.707$. Therefore, $\omega_n = \sqrt{2}/\tau_2 = 14.1$ rad/sec. (In terms of cycle bandwidth $B_L = 7.5$ Hz, and full IF bandwidth $= 38$ Hz.)

The Bode plot shows a phase margin of 30° and a gain margin of 6 dB. Although the loop is stable, its response will surely vary from that expected in the absence of the IF filter. If loop gain is fixed (by AGC or limiter) so that it cannot exceed the value used for the example, the stability margins are probably adequate but not ample. However, if the gain of the example is a threshold gain and increases of gain are to be expected with improved

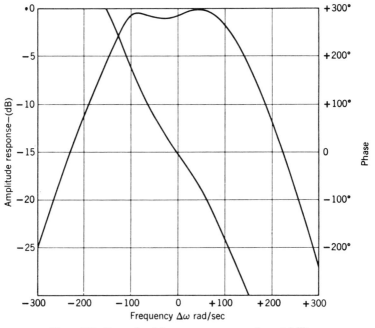

Figure 8.6 Example of frequency response of crystal filter.

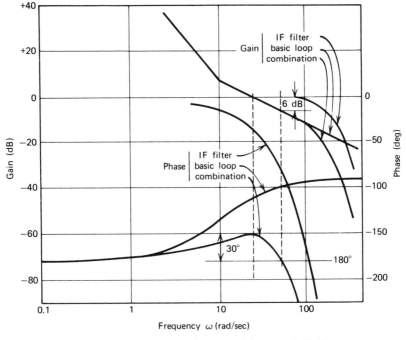

Figure 8.7 Bode plot of loop containing crystal IF filter.

signals, the gain margin is completely inadequate. If the gain doubles, the loop will oscillate. A more conservative design would use a substantially wider IF filter bandwidth.

False Locks

Even if the loop is stable, a narrow IF filter can cause acquisition difficulties in the form of *false locks*. Frequency search halts and the loop appears to lock at a frequency that bears no obvious relation to the input frequency. Until the source of false lock is recognized, the phenomenon can be a disturbing and mystifying experience.

It is shown shortly that false lock is an aberration of the pull-in mechanism and that it, or the related problem of *frequency pushing*, is almost inevitable in some degree in a PLL that includes extra filtering. Existence of false lock is another reason not to rely on pull-in as the frequency-acquisition method.

Other investigations of false lock have been performed by Develet[1], Johnson,[2] and Tausworthe.[3] The approximate analysis presented here follows a slightly different approach.

Let us consider an unlocked loop with input $V_s \sin \omega_i t$ and VCO output $V_o \cos \omega_o t$. Phase detector output is a beat-note at a frequency $\Delta \omega_i = \omega_i - \omega_o$. If $\Delta \omega_i$ is sufficiently larger than the loop gain K, the beat-note will be nearly sinusoidal and take the form $K_d \sin \Delta \omega_i t$.

In passing through the loop, the beat-note is attenuated by a factor $\eta(\Delta \omega_i)$ and phase shifted by an angle $\psi(\Delta \omega_i)$. Frequency-modulating voltage applied to the VCO is $\eta K_d \sin(\Delta \omega_i t + \psi)$, so the VCO output is (approximately)

$$v_o(t) = V_o \cos\left[\omega_o t - \frac{\eta K_o K_d}{\Delta \omega_i} \cos(\Delta \omega_i t + \psi)\right] \tag{8.1}$$

The spectrum (Figure 5.6) of $v_o(t)$ consists of a carrier line at ω_o and an infinite series of sideband lines at frequencies $\omega_o + k \Delta \omega_i$. The line for $k = 1$ is at a frequency of $\omega_o + (\omega_i - \omega_o) = \omega_i$, which is exactly the input frequency. Using a Fourier-series analysis, the VCO component at ω_i is found to be

$$V_o J_1\left[\frac{\eta K_o K_d}{\Delta \omega_i}\right] \sin(\omega_i t + \psi) \tag{8.2}$$

where $J_1(\cdot)$ is the first-order Bessel function of the first kind.

When this line is multiplied, in the phase detector, against the input signal $V_s \sin \omega_i t$, the resulting DC component is

$$V_d = \frac{1}{2} V_s V_o K_m J_1\left[\frac{\eta K_o K_d}{\Delta \omega_i}\right] \cos \psi = K_d J_1\left[\frac{\eta K_o K_d}{\Delta \omega_i}\right] \cos \psi \tag{8.3}$$

where K_m is as defined in Chapter 3.

In the standard second-order loop and for large enough $\Delta \omega_i$ we have $\eta = \tau_2/\tau_1$ and $\psi = 0$. Since $K_o K_d \tau_2/\tau_1 = K$, (8.3) becomes

$$V_d \simeq K_d J_1\left[\frac{K}{\Delta \omega_i}\right] \tag{8.4}$$

Equation 8.4 is an approximation to the pull-in voltage v_p of (5.6); the two expressions agree asymptotically for large frequency difference and disagree by less than 10% if $|\Delta \omega_i| > 2K$.

Now let us suppose that additional filtering is added into the standard loop. It is very difficult to avoid adding at least one extra pole for ripple filtering, the operational amplifier in an active filter contributes at least one more pole, and a third pole in the VCO control line is virtually inescapable. If a long loop is used, the filters in the IF amplifier contribute additional equivalent lowpass poles. Up to a dozen extra poles are not at all unusual.

We define a relative attenuation coefficient

$$\eta' = \frac{\eta K_o K_d}{K} \tag{8.5}$$

In the standard loop, $\eta' = 1$: departure of η' from unity describes the magnitude response of any additional filtering within a physical loop.

Accordingly, (8.4) is modified to

$$V_d = K_d J_1 \left[\frac{\eta' K}{\Delta \omega_i} \right] \cos \psi \tag{8.6}$$

The pull-in voltage of the standard loop (8.4) is multiplied by the cosine of the added phase shift. For $K/\Delta \omega_i \ll 1$ (the only region of validity for the approximations of this analysis) the Bessel function is approximated by

$$J_1 \left[\frac{\eta' K}{\Delta \omega_i} \right] \simeq \frac{1}{2} \left(\frac{\eta' K}{\Delta \omega_i} \right) \tag{8.7}$$

so the pull-in voltage is further reduced by a factor η'.

A suitable approximation for pull-in voltage, including the effects of additional filtering, is

$$V_d \simeq \frac{\eta' K_d K}{2 \Delta \omega_i} \cos \psi \tag{8.8}$$

If η' and ψ can be found, the pull-in and false-lock properties of the loop may be calculated from (8.8).

(Strictly speaking, the abbreviated analysis presented above applies directly only to a short loop. When the analysis is modified to take account of a long loop, the DC output of the PD can be estimated simply by cascading the equivalent modulation transfer function $F_m(s)$ with the actual loop filter $F(s)$ and calculating a new η' and ψ, provided that the bandpass amplifier is linear. If the bandpass circuit contains a limiter, the bandpass contribution to ψ is unaffected by the nonlinearity, but the contribution to η is more complicated. At large SNR the limiter tends to wipe off any influence on η contributed by bandpass networks preceding the limiter.)

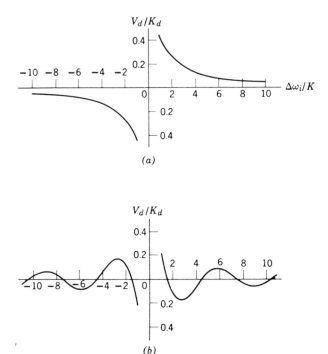

Figure 8.8 Loop pull-in characteristics showing effect of excess phase shift: (a) standard loop only; (b) excess phase $\psi = (\pi/3)(\Delta\omega_i/K)$.

As an example, we let phase be $\psi = (\pi/3)(\Delta\omega_i/K)$ and $\eta' = 1$, a fair approximation to the IF filter and PLL shown in Figures 8.6 and 8.7. Using this expression for ψ, the DC phase-detector output is plotted in Figure 8.8b.

Immediately evident in the plot are nulls of the pull-in voltage corresponding to the zeros of $\cos\psi$, nulls that do not occur in the standard loop (Figure 8.8a). For small $\Delta\omega_i$, the polarity of V_d is unchanged from that of the standard loop, so pull-in occurs correctly, albeit more weakly because of the reduced amplitude of V_d.

However, if the frequency difference is somewhat outside the first null the polarity of V_d is reversed from standard and pull-in no longer proceeds normally. Instead, the reversed polarity causes the loop to *push out* away from the correct lock frequency. Pushing continues until the frequency difference increases to coincide with the second null, which is a stable tracking point of *false lock*. Phaselock is certainly not achieved at the false-lock null, but the loop is unable to move itself away from false lock.

A false lock can be very confusing to an operator. Output from the loop phase detector will have zero DC component, whereas the quadrature PD (correlation detector) will show a DC output, indicating that lock has been achieved. If coherent AGC is used, the magnitude of the quad PD output is likely to be correct for indicating lock. An oscilloscope connected to the PD output will show the presence of a beat-note, but only if noise is small enough. In fact, it is possible that a false lock may go completely unrecognized—until post-fight data reduction comes up with some ridiculous Doppler shift.

Obviously, false locks should be avoided. One method of avoidance is to use an IF filter of sufficient bandwidth. Another is to recognize that phase shift, for a given bandwidth, increases as the number of resonant circuits in the filter increases. If only a single-tuned circuit is used, maximum phase shift is 90° and there is no finite spurious null. With two tanks (two poles in the equivalent low pass modulation transfer function) the maximum phase shift is 180° and the only finite spurious nulls are unstable.

Rough sketches of pull-in voltage for various numbers of poles are shown in Figure 8.9. Actual false locks are encountered only if there are four or more poles in the lowpass equivalent filter. Numerous poles are found in filters with very steep skirts—the so-called rectangular filters. Evidently such filters are not entirely suitable for use in a phaselock receiver.

A conservative design would utilize only one or two poles. (A single quartz crystal conveniently provides one equivalent pole.) Actually, there are certain to be other band-restricting elements within the loop, and there will always be more excess phase shift than is provided by the recognizable poles. The main IF filtering should be kept simple to provide some margin against these secondary effects, not all of which are easily predicted.

The generated DC voltage becomes very small as $|\Delta\omega_i|$ becomes large. As a result the effective phase-detector gain also becomes small. This trend is reinforced by the selectivity of the IF filter, which reduces the signal amplitude if $\Delta\omega_i \neq 0$. With reduced gain, the loop bandwidth is also smaller.

It is demonstrated in Chapter 4 that the maximum trackable frequency sweep rate depends on bandwidth; a narrowband loop can track only a slowly varying frequency. Therefore, if acquisition is performed by sweep techniques, it may be possible to sweep so rapidly that the false locks will be unable to hold, but, nevertheless, slowly enough to succeed in acquiring correct lock. This possibility is complicated by any limiters or AGC that may be used and by IF signal-to-noise ratio encountered.

The foregoing analysis takes into account only the normal signal path through the loop. Unfortunately, bitter experience has shown that insidious

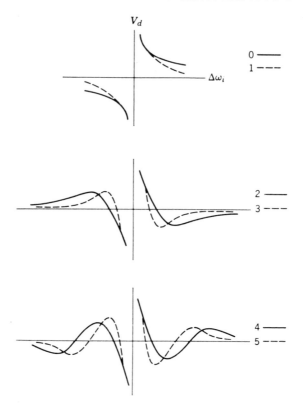

Figure 8.9 Loop pull-in characteristics. Numerals indicate equivalent number of extra lowpass poles in loop.

paths often contribute more to false lock than does the obvious main path. Beat-note coupling through an inadequately isolated power supply line is a particular offender.

Sideband Locks

Another improper lock is possible if the input signal is modulated by a periodic message waveform. This causes discrete sidebands to appear about the carrier frequency and the PLL may be able to lock to a sideband instead of the carrier. Sideband lock is particularly likely if the loop bandwidth is less than the sideband-to-carrier spacing.

Sideband lock is a normal attribute of a PLL and does not arise from a departure from normal, as does false lock. (There is no relation between sideband lock and false lock other than the fact that both are unwanted.)

It must be guarded against whenever there is a possibility of periodically modulated signals.

Anti-sideband-lock methods vary depending on the application; no generally valid remedy can be prescribed. Some workers[4,5] have used discriminator-aided acquisition to obtain correct frequency before phase-locking; this method is feasible only when the input SNR is reasonably large in the bandwidth of the discriminator. Some applications permit the modulation to be shut off until phase lock has been acquired. In other circumstances the carrier frequency is known with sufficient accuracy that the VCO can be set or swept within a range that includes the carrier but not the sidebands. Still other conditions might require a complicated, real-time analysis of the received signal to determine the correct lock line.

If the signal is amplitude modulated, a limiter removes the offending sidebands, leaving only the carrier for the loop to acquire.

8.2 AUTOMATIC GAIN CONTROL

A receiver designer almost always includes provisions for automatic gain control (AGC) in a receiver. Among the various reasons for such a practice are the following:

1. To control phaselock loop bandwidth.
2. To avoid overload.
3. To provide an indicator of signal level.
4. To aid in providing a reliable indicator of lock.
5. To standardize signal level in auxiliary channels (e.g., AM demodulators or antenna angle-tracking circuits).

The first reason is valid only in the absence of a limiter. If, as is common, a limiter is employed, it is the limiter that controls loop bandwidth, and the AGC (in most configurations) has no effect.

Avoidance of overload is probably the most important reason and is considered in some detail in this section.

Providing signal level indication may seem to be a trivial purpose, but a level indication is often needed, and measurement of AGC control voltage is a convenient way of obtaining it.

A loop lock indication is frequently needed for proper system operation, and use of AGC may be needed to obtain a reliable indication; the reason for this statement is examined briefly in later paragraphs.

We distinguish between "coherent" and "incoherent" (also called "wideband") AGC. The incoherent automatic gain control is derived conventionally by rectifying the output of an IF amplifier. Rectified voltage is proportional to the total output of the amplifier—the sum of signal, noise, and interference (if any). If noise exceeds the signal, as it may in typical phaselocked receivers, the rectified voltage is determined primarily by the noise.

Control voltage for coherent automatic gain control is obtained from a quadrature phase detector (QPD, see Figure 5.15) in which the DC component is proportional to signal alone. Noise causes fluctuations in the control voltage but contributes a negligible DC component. By suitable filtering after the phase detector the fluctuations can be reduced to any arbitrarily small level. As a consequence, the control voltage is proportional only to the signal and is independent of noise, even if noise at the PD input greatly exceeds the signal.

Overload

To explain the overload problem it is useful to have a hypothetical, though typical, receiver as an example. A double conversion receiver is shown in simplified block form in Figure 8.10 with three AGC pickoff locations indicated. Two are incoherent and the third is coherent. Note that AGC must be generated before the limiter and not after.

To be able to assign numbers to the example, we suppose that receiver noise figure is 10 dB so that receiver noise power is -184 dBW in the bandwidth $2B_L = 10$ Hz. Using (3.21) and assuming that the receiver threshold occurs when signal-to-noise ratio in the loop is 0 dB, we calculate the threshold sensitivity to be -184 dBW. Coherent AGC can be made operative at any signal level for which the loop locks.

At receiver threshold the output signal from the second IF amplifier is 23 dB below the noise level; from the first IF amplifier it is 50 dB below

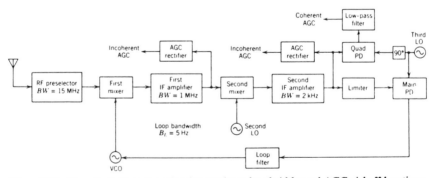

Figure 8.10 Hypothetical receiver showing pertinent bandwidths and AGC pickoff locations.

the noise level. These outputs obviously cannot provide signal-related, rectified voltages under weak signal conditions. For a wide range of useful signals conventional incoherent AGC is determined solely by noise. To obtain signal-related gain control, coherent detection is essential.

If only coherent AGC is used, no AGC is generated until the receiver has acquired lock. Once lock has been obtained, the likelihood of signal overload is small, but until then it is possible that the receiver might be overloaded. To prevent strong signal overload from occurring before lock-up, designers sometimes provide two-mode AGC: both coherent and incoherent AGC detectors. Actual control is exercised by whichever detector generates the strongest output voltage. In the unlocked condition only the incoherent detector could possibly have control, whereas the coherent detector would take over once lock had been established.

It has already been indicated several times that limiters are commonly used in phaselocked receivers. A limiter is constantly in an overloaded condition; therefore, why should there be concern if stages preceding the limiter also overload? The answer lies in the detailed design of any particular receiver. If, in fact, the various stages act as good limiters when they are overloaded, we need not be unduly concerned with strong signal overload; the receiver would probably tolerate it.

Difficulties arise, however, because little or no thought is ordinarily given to the nature of overload behavior in the design of IF amplifiers. Rather than limit cleanly, an amplifier might very well oscillate, block, squeg, or detune when heavily overloaded. As a consequence, the signal would be damaged, and severe lock-up difficulties might be experienced. The problem can be avoided by suitable design of the individual IF stages, but it is also avoided by a two-mode AGC. Of the two solutions, the second is often the simpler.

Under conditions of very weak signal there is no signal-caused overload, but the noise exceeds the signal, with the resulting possibility of noise overload. In a sensitive receiver, it is usual that the largest voltages to be handled by the IF amplifiers occur under the weakest signal conditions. The signal is deeply buried in noise; since the coherent AGC holds signal to a standard level, the noise grows very large. Circuits intended to be linear must be designed not to saturate on the weak-signal noise.

Lock Indication

A positive indication of lock is often required in a receiver. In principle, AGC has nothing to do with any lock indicator—an indication could be obtained from a quadrature phase detector operating on the same signal as the loop phase detector. In most receiver configurations, however, the signal level (without AGC) at the detector would be variable over a wide range as the input signal varied. Because output from the quadrature PD is proportional to signal, the magnitude of the lock indication voltage would be highly variable.

The DC output of the QPD must be applied to some decision device that provides a "locked" or "unlocked" message to the outside world. Such a device is difficult to build if the decision level has a widely variable magnitude.

If coherent AGC is used, the problem disappears. Operation of the AGC is such as to maintain DC output of the QPD constant for all signals, as long as the receiver is locked. In the unlock condition, the DC output is zero (or at least very small). A fixed decision level is entirely adequate, and no dynamic range problems develop.

The IF amplifiers driving the QPD and the QPD itself must not overload if the level indication is to be correct. Since, in a typical receiver, the noise level can be substantially larger than a weak signal, the signal level delivered to the QPD will probably have to be fairly small, or else noise will overload the amplifier. Further amplification (DC) following the QPD and filter is often needed.

AGC Bandwidth

Speed of response of the AGC loop is a question that has not been explored in any depth in the existing literature. For deep-space missions where signal strength changes very slowly, it has been the practice to build AGC with a very slow response[4]. Typically, the AGC noise bandwidth is much narrower than the phaselock loop bandwidth (by as much as 1000 to 1). With such a great disparity of bandwidths, noise fluctuations remaining in the AGC loop have a negligible effect on the tracking loop.

There are situations in which signal level can change rapidly and drastically. If AGC response is too slow, the receiver gain does not compensate quickly enough, and a signal reduction appears, in effect, to be a complete dropout. As a result, the receiver loses lock. Obviously, the AGC response must be fast enough to follow any large variations of the signal.

If AGC response is fast, its noise bandwidth must be correspondingly wide, and the receiver gain therefore fluctuates with the noise. One would expect these gain fluctuations to have some effect on phaselock tracking, but there has been no analysis to predict the nature of the effect. Some fragmentary experimental evidence suggests that the reaction on the tracking loop is small; however, more work is needed.

A word on stability of AGC loops is also in order. If a narrow IF filter, with bandwidth comparable to the AGC loop bandwidth, is included within the AGC loop, it is entirely likely that the loop will oscillate. The problem is similar to oscillation of the tracking loop caused by the same phenomena, as is explained earlier in this chapter.

8.3 COHERENT TRANSPONDERS

A *transponder* receives a signal, processes it in some manner, and retransmits the signal at increased power. A transponder is said to be *coherent* if its transmitted frequency f_t is a rational multiple of its received frequency f_r; that is, $f_t = (m/n)f_r$, where m and n must be integers. With this definition of coherence, there are exactly m cycles out for every n cycles that enter the transponder. The frequency received at the ground can be multiplied by n/m and the result can be compared against the frequency originally transmitted from the ground; their difference is the two-way Doppler shift.

Early coherent transponders often used $n = 1$; thus the output frequency was a harmonic (usually the second) of the input. A transponder of this type need not be phaselocked to be coherent. Our interest here is in an offset transponder in which neither m nor n is unity. The output frequency is offset—usually by a relatively small amount—from the input frequency. Coherence in offset transponders is almost always obtained by means of phaselock techniques.

A block diagram of a typical phaselocked transponder is shown in Figure 8.11. Double superheterodyne conversion is illustrated in the receiver portion, but single or triple conversion receivers operate on the same principles. All mixer and phase-detector injection voltages are obtained as harmonics of a single local oscillator, and the output frequency is also a harmonic of the same oscillator. It is now shown that if the loop is locked the output will be coherent with the input.

Operation of the first mixer may be described by the equation

$$f_r = N_1 f_o \pm f_1 \tag{8.9}$$

operation of the second mixer is given by,

$$f_1 = N_2 f_o \pm f_2 \tag{8.10}$$

and the phaselock requirement is

$$f_2 = N_3 f_o \tag{8.11}$$

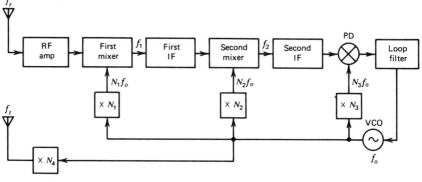

Figure 8.11   Phaselock transponder.

where nomenclature is defined in Figure 8.11 and the choice of plus or minus signs depends on whether low-side or high-side injection, respectively, is used. A combination of these three equations and elimination of the two intermediate frequencies results in

$$f_r = f_o(N_1 \pm N_2 \pm N_3)$$

Because the transmitted frequency is $f_t = N_4 f_o$, the ratio of output frequency to input frequency is

$$\frac{f_t}{f_r} = \frac{N_4}{N_1 \pm N_2 \pm N_3} \qquad (8.12)$$

which is a rational number. Therefore, by our previous definition, the transponder is coherent if it is locked.

In a practical transponder the multiplication ratios are often chosen with the result that N_1, N_2, and N_4 have many common prime factors that permit the individual frequency multipliers to be combined to a substantial degree. The three individual multipliers tend to coalesce into one string of multipliers with three output taps.

It is common practice to set $N_3 = 0.5$ and use a frequency divider instead of a multiplier at this location. A parametric divider[5], a regenerative divider[6,7], or, if the VCO frequency is low enough, a binary counter,* is often employed. At first appearance such design seems rather odd—why

---

*Binary dividers have fast transitions that are likely to induce spikes onto power and ground lines. The spike rate or a harmonic thereof is likely to fall at one of the internal frequencies and therefore cause unacceptable interference in a sensitive receiver.

use a divider? The reason is clear if we suppose that a multiplier is used instead; in particular, assume that $N_3 = 1$, so that the VCO and second IF are at the same frequency. Any physical multiplier has some output at its fundamental frequency as well as at its desired harmonic, and the output of the N_2 multiplier contains a small component at frequency f_o. However, that component is in the center of the second IF passband and is strongly amplified by the second IF amplifier. This feed-around signal will surely interfere with proper operation of the loop, and it is entirely possible that the loop will actually lock to itself. If we make $N_3 = 0.5$ and provide large reverse-direction attenuation between divider and VCO, no component of the second IF will be able to loop around and cause self-lock.

The various multiplier factors and the overall translation ratio must be chosen with strong consideration given to possible interference effects that might arise from harmonics, mixer products, and so on. A widely used ratio is $m/n = 240:221$; one way to achieve that ratio is to set $N_1 = 108$, $N_2 = 3$, $N_3 = 0.5$, and $N_4 = 120$.

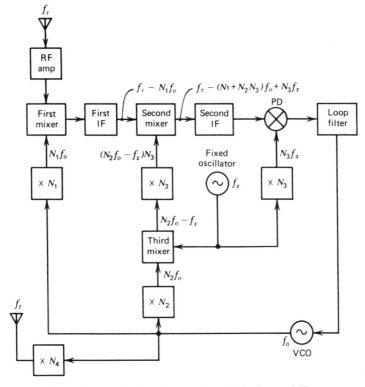

Figure 8.12 Phaselock transponder with fixed second IF.

The intermediate frequencies in Figure 8.11 are not fixed; instead, they vary as the input frequency varies, although over smaller limits. Phase-slope of the IF filters must be small enough so that the changing frequency will not produce a significant Doppler error.

The second intermediate frequency is held fixed in the configuration of Figure 8.12. Phaselock imposes $f_o = f_r/(N_1 + N_2 N_3)$, independent of the fixed frequency f_x. The fixed oscillator is incoherent with the incoming signal, the first IF, or the VCO; f_x is cancelled by adding $N_3 f_x$ into the second mixer and subtracting $N_3 f_x$ at the phase detector. An arrangement of this kind allows greater flexibility in selection of internal frequencies. For the same reasons as before, the third multiplier is usually chosen as $N_3 = 0.5$.

REFERENCES

1. J. A. Develet, Jr., "The Influence of Time Delay on Second-Order Phase Lock Loop Acquisition Range," *Int. Telem. Conf.*, pp. 432–437, London, 1963.

2. W. A. Johnson, *A General Analysis of the False-Lock Problem Associated with the Phase-Lock Loop.* Report TDR-269 (4250-45)-1, Aerospace Corp., October 2, 1963 (NASA Accession N64-13776).

3. R. C. Tausworthe, "Acquisition and False-Lock Behavior of Phase-Locked Loops with Noisy Inputs," *JPL SPS 37-46*, Vol. IV, pp. 226–234, Jet Propulsion Laboratory, Pasadena, CA, August 31, 1967.

4. W. K. Victor and M. H. Brockman, "The Application of Linear Servo Theory to the Design of AGC Loops," *Proc. IRE*, Vol. 48, pp. 234–238, February 1960.

5. R. A. Mostrom, "The Charge-Storage Diode as a Subharmonic Generator," *Proc. IEEE*, Vol. 53, p. 735, July 1965.

6. C. W. Helstrom, "Transient Analysis of Regenerative Frequency Dividers," *IEEE Trans.*, CT-12, pp. 489–497, December 1965.

7. S. V. Ahamed, J. C. Irvin, and H. Seidel, "Study and Fabrication of a Frequency Divider–Multiplier Scheme for High-Efficiency Microwave Power," *IEEE Trans.*, COM-24, pp. 243–249, February 1976.

Chapter Nine

Phaselocked Modulators
and Demodulators

Phaselocked demodulators for amplitude modulation (AM), phase modulation (PM), and frequency modulation (FM) are widely used. Coherent demodulation of AM or PM is almost always accomplished with the help of a phaselocked loop; phaselocked FM demodulators can achieve lower thresholds than conventional FM discriminators.

Angle modulators (PM and FM) are sometimes mechanized by means of phaselocked loops. Transfer functions and performance features are presented here.

The PLLs treated in this chapter are used mainly for processing analog signals, although they could also handle some forms of digital modulation. For descriptions of PLLs that are specialized for digital signals, see Chapter 11.

9.1 PHASELOCKED MODULATORS

There are numerous methods of producing phase modulation or frequency modulation. In one method the baseband message is inserted into the low-frequency portion of a PLL so as to phase modulate or frequency modulate the VCO.* Center-frequency stability is established by the fixed oscillator that serves as a stable reference. Locking forces the average VCO frequency to be equal to the reference frequency. The loop tracks out frequency drift of the VCO.

---

*The distinction between PM and FM is artificial; both might be termed *angle modulation* and treated in a unified manner. In this chapter, the term "PM" implies small phase deviation, with a remanent carrier present, whereas "FM" has no such implications. The distinction is more apparent in the modulator and demodulator configurations than in the signals themselves.

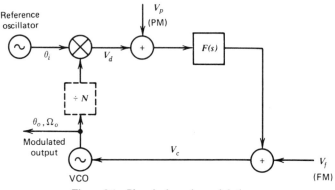

Figure 9.1 Phaselock angle modulation.

A block diagram of an angle-modulated PLL is shown in Figure 9.1. Phase modulation is accomplished by adding the modulating voltage V_p to the output of the phase detector V_d. The loop attempts to maintain the sum $V_p + V_d$ at a null; this is possible only if a phase error generates a V_d that cancels V_p. Phase error arises from phase modulation of the VCO, which is the purpose of the circuit.

By use of the transfer-function methods of Chapter 2, the VCO phase modulation caused by the voltage V_p is found to be*

$$\theta_o(s) = \frac{K_o F(s) V_p(s)}{s + K_o K_d F(s)} = \frac{V_p(s)}{K_d} H(s) \tag{9.1}$$

Modulation sensitivity of the circuit is $1/K_d$ rad/V. Since $H(s)$ is a lowpass function, the loop bandwidth must be larger than the highest modulation frequency to avoid linear distortion.

The phase detector characteristic must be linear to avoid nonlinear distortion of the modulation. Nonlinearity of the VCO is reduced by feedback; VCO nonlinearity is tolerable if loop bandwidth is sufficiently larger than modulating frequency.

Frequency modulation is produced by adding a baseband voltage V_f into the VCO control terminal along with the output of the loop filter. Output phase modulation is readily shown to be

$$\theta_o(s) = \frac{K_o V_f(s)}{s + K_o K_d F(s)} \tag{9.2}$$

*From Figure 9.1, reference phase θ_i is assumed constant and may be ignored. Also, we temporarily assume $N = 1$.

Output frequency modulation is the derivative of the phase:

$$\Omega_o(t) = \frac{d\theta_o(t)}{dt} \tag{9.3}$$

which, when L-transformed, becomes

$$\Omega_o(s) = s\theta_o(s) \tag{9.4}$$

Therefore, the transfer function from V_f to Ω_o is

$$\Omega_o(s) = \frac{sK_oV_f(s)}{s + K_oK_dF(s)} = K_oV_f[1 - H(s)] \tag{9.5}$$

Since $1 - H(s)$ is a highpass function, the loop bandwidth (highpass corner frequency) must be smaller than the lowest modulation frequency. The phaselocked frequency modulator cannot generate a constant frequency offset.

To avoid nonlinear distortion, the VCO control characteristic must be linear. Feedback compensates for nonlinearity of the PD characteristic.

Output phase deviation (for either PM or FM) appears as a phase error at the phase detector. Since the phase detector has limited range, it is not possible to obtain a large modulation index (peak phase deviation) with the circuit in solid blocks in Figure 9.1. Extended-range PDs (see Chapter 6) are of some help, but the best of these detectors restricts the phase deviation to less than 2π rad, peak.

To achieve a larger modulation index, one could operate the VCO at an integer harmonic N of the input reference frequency and divide the VCO frequency by N (dashed block in Figure 9.1) before applying it to the phase detector. In this manner the peak phase error is $1/N$ times the peak deviation. Arbitrarily large indices can be generated by making N sufficiently large.

9.2 PHASELOCKED DEMODULATORS

Phaselocked loops are used for demodulation of many kinds of modulated signals. Applications include:

- Coherent amplitude detectors (product detectors).
- Phase demodulators (PM detectors).
- Frequency demodulators (FM discriminators).

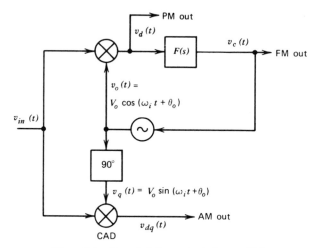

Figure 9.2 Demodulation outputs from a PLL.

Figure 9.2 shows pickoff points in a PLL for recovery of each type of modulation and establishes nomenclature for the discussion to follow.

PLL Response to AM

We let the input signal be amplitude modulated, that is,

$$v_{in}(t) = V_s x(t) \sin(\omega_i t + \theta_i) \tag{9.6}$$

where $x(t)$ is arbitrary, dimensionless amplitude modulation and the other symbols are the same as in Chapter 3.

The phase detector is conveniently modeled as a multiplier that generates the product of v_{in} and the VCO output. Discarding double-frequency terms, the PD output is found to be

$$v_d(t) = K_d x(t) \sin \theta_e \tag{9.7}$$

Only the average (DC) output of the phase detector is useful in establishing phaselock; any fluctuating components are just potential sources of tracking disturbance. The average value of V_d is

$$\text{avg}[v_d(t)] = \text{avg}[x(t)] K_d \sin \theta_e \tag{9.8}$$

There is useful output—the loop is able to lock—only if $\bar{x} \triangleq \text{avg}[x(t)] \neq 0$.

The modulation must have a DC component for a discrete carrier component to be present. An ordinary PLL needs a carrier component to which it can lock. Conversely, an ordinary PLL is unable to lock to a suppressed-carrier signal. (See Chapter 11 for methods of nonlinear regeneration of a carrier from a signal that has none.)

We represent the modulation as $x(t) = x'(t) + \bar{x}$, where x' has zero mean and $\bar{x} \neq 0$. Phase detector output becomes

$$v_d(t) = [x'(t) + \bar{x}] K_d \sin \theta_e \tag{9.9}$$

But, if the loop is phaselocked and is tracking properly, then $\theta_e \approx 0$ and there is near-zero output from the phase detector, irrespective of the properties of $x'(t)$. Therefore, to a first approximation, a PLL does not respond to AM that might be present on its input.

(For a more concrete example, let us consider sinusoidal AM applied to a perfectly tuned first-order PLL, so the loop equations are, discarding double-frequency terms:

$$v_{in}(t) = V_s(1 + m \sin \omega_m t) \sin(\omega_i t + \theta_i)$$

$$v_d(t) = K_d(1 + m \sin \omega_m t) \sin(\theta_i - \theta_o)$$

$$\frac{d\theta_o}{dt} = K_o v_d$$

We let $\theta_i = 0$ and recollect that $K_o K_d = K$. Combining the above equations yields the differential equation of the loop

$$\frac{d\theta_o}{dt} = -K(1 + m \sin \omega_m t) \sin \theta_o$$

or rearranging

$$\frac{d\theta_o}{\sin \theta_o} = -K(1 + m \sin \omega_m t) dt$$

Integrating both sides gives

$$\ln\left[\tan \frac{\theta_o}{2}\right] = -Kt + \left(\frac{mK}{\omega_m}\right) \cos \omega_m t + C$$

where C is a constant of integration. Taking exponentials

$$\tan\frac{\theta_o}{2} = \exp(-Kt)\exp\left(\frac{mK}{\omega_m}\cos\omega_m t\right)\exp C$$

which vanishes for large t.

Therefore, if a PLL ultimately tracks with zero phase error, the presence of amplitude modulation does not alter the equilibrium condition nor does it introduce any phase modulation of the VCO. If steady-state phase error is not zero, there is a complicated, nonlinear interaction between modulation and phase error that arises again when we consider FM demodulators.)

Coherent Amplitude Detector

Following the development of Chapter 3, we consider that the input to the PLL of Figure 9.2 consists of an amplitude-modulated signal plus additive, narrowband, gaussian noise:

$$v_{in}(t) = x(t)V_s\sin(\omega_i t + \theta_i) + n_c(t)\cos(\omega_i t + \theta_i) - n_s(t)\sin(\omega_i t + \theta_i)$$

$$(9.10)$$

We multiply the input by a 90° phase-shifted version of the VCO output (Figure 9.2)

$$v_q(t) = V_o\sin(\omega_i t + \theta_o) \tag{9.11}$$

whereupon the difference-frequency output of the multiplier is

$$v_{dq}(t) = K_d\left[x(t)\cos\theta_e + \frac{n_c(t)}{V_s}\sin\theta_e - \frac{n_s(t)}{V_s}\cos\theta_e\right] \tag{9.12}$$

where $\theta_e = \theta_i - \theta_o$ and K_d is as defined in Chapter 3.

If the VCO is tracking properly, then θ_e is nearly zero, so the output of the coherent amplitude detector (CAD) is, closely,

$$v_{dq}(t) \approx K_d\left[x(t) - \frac{n_s(t)}{V_s}\right] \tag{9.13}$$

which consists of the linear sum of the desired amplitude modulation plus the component of noise modulation that lies in phase with the signal. The quadrature noise component and moderate amounts of phase modulation are rejected. The CAD performs amplitude demodulation.

Amplitude demodulation could also be performed by a simple envelope detector, provided that $x(t)$ is always positive. If $x(t)$ becomes negative ("overmodulation" in radio-engineering parlance), then an envelope detector generates severe distortion.

The coherent amplitude detector imposes no such restriction; it reproduces $x(t)$ without distortion, even if $x(t)$ reverses polarity. Moreover, the CAD will even demodulate a suppressed-carrier signal, provided there is some means of generating a properly phased local reference—some means of locking the VCO to the proper phase (see Chapter 11.)

Coherent amplitude detectors are also used for low-distortion demodulation of single-sideband (SSB) and vestigial sideband (VSB) signals.[1] A remanent pilot carrier must be transmitted with the signal if the local carrier reference is to be phaselocked (as is essential for coherent demodulation of VSB).

A major advantage of a CAD lies in its linear processing of signal and noise. Its output is the linear superposition of signal and noise, irrespective of the input signal-to-noise ratio; there is no intermodulation between the two.

From (9.10) the input signal-to-noise ratio is (remembering that $\overline{n_c^2} = \overline{n_s^2}$)

$$\text{SNR}_i = \frac{\frac{1}{2} V_s^2 \overline{x^2(t)}}{\overline{n_s^2(t)}} \tag{9.14}$$

while (9.13) shows the output SNR to be

$$\text{SNR}_o = \frac{V_s^2 \overline{x^2(t)}}{\overline{n_s^2(t)}} = 2\text{SNR}_i \tag{9.15}$$

Equation (9.15) holds for all values of SNR_i.

In consequence of the linear processing of noise, it is equally effective to filter out noise, either prior to the CAD by means of a bandpass filter or after it by means of a baseband filter. Since a baseband filter is typically much easier to implement than the equivalent bandpass filter, that option is often extremely valuable to a designer.

Predetection and postdetection filtering are equally effective because the CAD does not degrade the signal-to-noise ratio. By contrast, an envelope detector causes a characteristic *squaring loss* for small SNR_i. Under conditions of low input SNR the output is proportional to $(\text{SNR}_i)^2$; for each 1-dB loss at the input, there is a 2-dB loss at the output of the envelope detector.[2]

If SNR_i is appreciably larger than unity, the CAD and the envelope detector have the same SNR_o. At small SNR_i, use of an envelope detector imposes a substantial penalty. See Figure 9.3 for a comparison of the performances of the coherent and incoherent detectors.

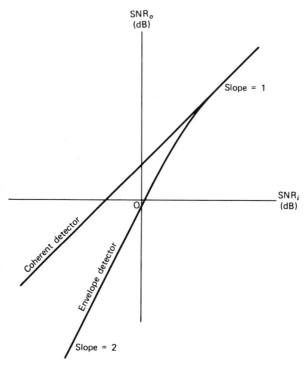

Figure 9.3 AM detector comparison. Vertical distance between curves is squaring loss in decibels.

Phase Demodulation

We assume that the input signal is phase modulated:

$$v_{in}(t) = V_s \sin[\omega_i t + \theta_i(t)] \qquad (9.16)$$

where $\theta_i(t)$ is the phase modulation. If the peak phase excursion is small enough that the PLL remains in its linear domain, then the linear transfer-function analysis of (2.3) and (2.7) applies and the output of the phase detector can be represented as

$$V_d(s) = K_d \theta_i(s) \left[\frac{s}{s + K_o K_d F(s)} \right] \qquad (9.17)$$

The bracketed quantity can be shown to have the characteristics of a highpass filter (see Figure 2.4 for an example) whose value approaches unity for large values of $s = j\omega$. In other words, phase modulation of sufficiently high frequency appears, unaltered, at the output of the loop phase detector.

Origin of the highpass action is readily understood from physical reasoning. The loop tracks low-frequency modulation and fails to track modulation frequencies that are outside its bandwidth; it is a *carrier-tracking loop*, as defined in Chapter 4. Output of the phase detector is a measure of the untracked phase error, so high-frequency modulation components appear there, unchanged by the loop dynamics. A lower-frequency component is reduced by the feedback factor at that frequency.

Carrier-tracking coherent PM systems must be designed so that the demodulator loop does not suppress the modulation. Subcarriers are often employed to move the information spectrum outside the PLL bandwidth. A subcarrier also moves the signal information away from low-frequency noise and drift disturbances of the VCO. See Chapter 6.

Undistorted demodulation is achieved if the peak phase excursion remains within the linear portion of the phase-detector characteristic. To enhance linearity, an extended-range detector, as is described in Chapter 6, might sometimes be useful. However, since all PD characteristics revert to sinusoidal for low SNR_i, the sinusoidal characteristic is of great importance.

Distortion can be tolerated in some applications; demodulation to a subcarrier might be one example. The phaselock loop demands that the signal contain a trackable carrier, but the modulation index is otherwise unrestricted.

We assume that the modulation is sinusoidal with a modulating frequency ω_m that is outside the bandwidth of the loop. Modulated input phase is

$$\theta_i(t) = \Delta\theta \sin \omega_m t \tag{9.18}$$

The loop is unable to track the modulation, so phase error $\theta_e = \theta_i$ and the output of the phase detector is

$$v_d(t) = K_d \sin(\Delta\theta \sin \omega_m t) \tag{9.19}$$

which is a nonlinear function of the modulation.

Some examples of distorted output waveforms are given in Figure 9.4 for various choices of peak deviation $\Delta\theta$. Distortion clearly worsens as $\Delta\theta$ becomes larger. The plots all are for a zero mean value of θ_e; any phase offset would cause asymmetric distortion.

since frequency is simply the derivative of phase. Taking Laplace transforms we obtain

$$L\{m(t)\} = M(s) = s\theta_i(s) \tag{9.21}$$

and substituting into (9.20) gives

$$V_c(s) = \frac{M(s)H(s)}{K_o} \tag{9.22}$$

which shows the transfer function between frequency modulation and the resulting VCO control voltage. The recovered message is equivalent to the original message, filtered by the closed-loop transfer function $H(s)$ and scaled by the VCO gain factor K_o. If the loop is linear and if its bandwidth is large enough compared to the message bandwidth, $v_c(t)$ is a faithful reproduction of $m(t)$.

To avoid distortion, it is evident that the VCO control characteristic must be linear, since K_o appears directly in (9.22); that is, K_o must truly be a constant and not a function of v_c.

Phase-detector gain enters (9.22) only through its influence on $H(s)$—which is significant only at higher modulation frequencies, since $H(0) = 1$ irrespective of K_d. For this reason, and because of the reduction of PD distortion by feedback that is noted in Chapter 4, low-distortion operation is possible with a nonlinear phase detector.

Avoidance of linear filtering distortion, avoidance of nonlinear PD distortion[6], and, indeed, the very ability to maintain track (Chapter 4) are all enhanced by large bandwidth of the PLL. These interrelated reasons all point to a loop bandwidth that is much larger than the message bandwidth. It becomes apparent later that the loop bandwidth should actually be substantially larger than the RF bandwidth of the modulated signal, a conclusion that is not obvious at this point.

FM Noise

We let the frequency modulation be sinusoidal with peak deviation Δf Hz and modulating frequency f_m Hz. Therefore,

$$m(t) = 2\pi\Delta f \sin 2\pi f_m t \tag{9.23}$$

and the input signal becomes

$$v_{in}(t) = V_s \sin\left(\omega_i t + \frac{\Delta f}{f_m} \cos 2\pi f_m t\right) \tag{9.24}$$

so the PLD output signal is

$$v_c(t) = \frac{1}{K_o} 2\pi\Delta f \sin 2\pi f_m t \qquad (9.25)$$

where it has been assumed that $H(j2\pi f_m) \simeq 1$; that is, the loop does not filter the modulation appreciably.

Preceding the demodulator is a bandpass filter centered at the signal frequency ω_i and with noise bandwidth B_i Hz. The filter is assumed to have sufficient bandwidth, amplitude flatness, and phase linearity to cause negligible distortion to the signal. A lower-bound constraint is $B_i > 2\Delta f$ and practical bandwidths are usually substantially larger.

White, gaussian noise of one-sided density N_o V^2/Hz is added to the signal at the filter input. Signal-to-noise power ratio at the filter output is

$$\rho_i = \frac{V_s^2}{2B_i N_o} \qquad (9.26)$$

This quantity is the so-called *carrier-to-noise ratio* (CNR) that appears throughout the FM literature.

The effect of noise in a PLL is represented in Chapter 3 as an additive noise generator $n'(t)$, with spectral density $\Phi_{n'}(\omega)$ rad^2/Hz, inserted into the linearized phase detector (Figure 3.2 and equations 3.6 and 3.13). If the bandpass filter has a rectangular passband (not a necessary assumption) and $\rho_i \gg 1$ (necessary to assure linearity) then $\Phi_{n'}(\omega) = 2N_o/V_s^2$ for $0 \leqslant \omega \leqslant \pi B_i$ and is zero otherwise.

By use of the transfer-function methods of Chapters 2 and 3, the spectral density of noise appearing in the control voltage v_c is found to be

$$\Phi_c(\omega) = \left[\omega^2 \Phi_{n'}(\omega) \right] \frac{|H(j\omega)|^2}{K_o^2} \qquad \text{V}^2/\text{Hz} \qquad (9.27)$$

If $\Phi_{n'}$ is flat, the bracketed term has the familiar parabolic spectrum associated with FM noise.

Signal and noise of the control voltage are processed through an external lowpass post filter. We assume the passband to be rectangular with cutoff frequency equal to the modulating frequency f_m. The recovered modulation is passed without loss, but all higher-frequency components of noise are suppressed completely.

We assume that $|H(j2\pi f)|$ is flat from DC to f_m. Noise intensity at the output of the post filter is given by

$$\sigma^2 = \frac{1}{2\pi} \int_0^{2\pi f_m} \Phi_c(\omega)\, d\omega = \frac{1}{2\pi} \left[\frac{2N_o}{V_s^2 K_o^2} \right] \frac{(2\pi f_m)^3}{3} \qquad \text{V}^2 \qquad (9.28)$$

Output signal-to-noise ratio is given by the mean square of (9.25) divided by (9.28) or

$$SNR_o = \frac{3\Delta f^2 V_s^2}{4N_o f_m^3} = \frac{3\Delta f^2 B_i \rho_i}{2f_m^3} \qquad (9.29)$$

If $\Delta f = B_i/2$ (the maximum deviation that remains within the input filter) and we define $\beta = \Delta f / f_m$, then

$$SNR_o = 3\beta^3 \rho_i \qquad (9.30)$$

which is the classical expression of *FM improvement factor*.[7,8] This result is exactly the same as that obtained for a conventional frequency discriminator. For large CNR a PLD has noise performance identical to that of ordinary discriminator circuits.

To achieve the FM improvement, a conventional discriminator must be preceded by a limiter. Ordinary discriminator circuits are amplitude sensitive and the limiter is essential to suppress the AM component of noise.

The PLD furnishes the FM improvement without employing a limiter. In effect, the PLL ignores the component of noise that lies in phase with the signal and is disturbed only by the quadrature component. It is demonstrated in the next section that a limiter worsens threshold performance; ability to deliver FM improvement without incurring limiter loss is one motive for using a phaselocked discriminator in place of a conventional circuit.

9.3 FM THRESHOLD

The ideal performance of (9.29) is achieved at high CNR, but below some minimum CNR—known as the *threshold*—the output SNR_o deteriorates very rapidly with further reduction of CNR. This section is devoted to an exposition of the threshold effect.

The reader should be forewarned that no good quantitative theory has yet been devised for the explication of threshold of a PLD. Operation of the loop is in the nonlinear region, but the nonlinear methods reported in Chapter 3 are inadequate to cope with a PLL subjected to bandlimited additive noise simultaneously with modulation. (Existing Fokker-Planck methods are restricted to treatment of white input noise.)

Here we attempt a heuristic explanation of PLD threshold based on experimental evidence and laboratory experience. The explanations are inadequate in that the threshold CNR cannot be predicted nor can the optimum loop configuration be calculated. However, sufficient information is given so that an engineer can optimize design parameters by experiment.

Threshold Characterization

Excessive deterioration of output SNR_o is the best-recognized manifestation of FM threshold, as sketched in Figure 9.6. At high CNR the output SNR_o is linearly proportional to CNR, as per (9.29). [Although (9.29) was derived for sinusoidal modulation, similar expressions can be obtained for any other modulation format.] The SNR_o versus CNR curve has unit slope on log–log coordinates for large CNR.

At threshold CNR there is a break in the slope, and the curve is much steeper for low CNR. Since the curve is continuous, an exact break point is difficult to recognize. It is usual to take the point of 1-dB deterioration from the extended straight line as the formal definition of threshold CNR, but that choice is entirely arbitrary.

Threshold performance of an *ideal* frequency discriminator is taken as a reference against which to compare other devices. An ideal discriminator produces an output baseband voltage that corresponds to the rate of

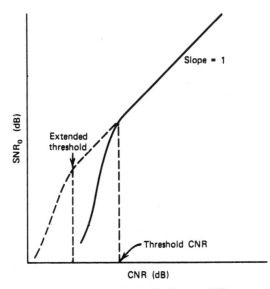

Figure 9.6 FM threshold effect on SNR_o.

change of phase of the bandpass process applied to its input: the instantaneous frequency of signal-plus-noise. The phase is that of the resultant of the desired signal plus added noise. An ideal discriminator is insensitive to AM components of signal or noise; performance of well-designed, conventional limiter–discriminator circuits presumably is very close to ideal.

The term "ideal" in no way implies "optimum" in this context. All good discriminators have the same performance at large CNR, but the ideal discriminator does not have the lowest threshold. If a discriminator has a lower threshold than ideal, it is said to be an *extended-threshold* demodulator. An example of threshold extension is sketched in Figure 9.6.

A phaselocked demodulator is valuable because it offers threshold extension, as compared to an ideal discriminator, with a relatively simple circuit. The amount of extension is not predictable by any existing theory and depends on signal parameters; very roughly, a few decibels of improvement are achieved in typical applications.

Output SNR_o of an ideal discriminator can be calculated from an exact analysis[9,10] by Shimbo that was produced after a long series of approximate analyses by earlier authors. Difficulty of the problem is perhaps best illustrated by the fact that the exact analysis was not published until some 45 years after the nature of FM was recognized.[11]

In many applications the SNR_o below threshold is of little interest because normal operation occurs almost exclusively at CNRs above threshold. (The disturbances accompanying below-threshold operation are often much more disruptive than might be expected from SNR_o considerations alone; their nature is described shortly.) It is often sufficient, for signal design and link budget purposes, to be able to predict the threshold CNR.

The prediction of threshold could, of course, be performed by evaluation of Shimbo's equations, but an approximate method by Rice[8] is easier to use and also has concepts that aid understanding of the PLD.

Output of a below-threshold discriminator can be observed to contain large-amplitude, short-duration spikes or *clicks* (to use Rice's term). These clicks appear only very rarely above threshold. More-frequent appearance of clicks is a manifestation of the onset of threshold.

Note the wording of the last sentence. It is not stated that clicks cause threshold or that clicks are the only manifestation of below-threshold operation; neither would be true. Nonetheless, the threshold CNR can be predicted with good accuracy if the average click rate can be calculated.

A click results when noise causes the resultant of signal plus noise to take on (or lose) one complete cycle as compared to the signal alone. A phaser diagram (Figure 9.7) illustrates the generation of clicks. The phaser reference is chosen so that the signal remains fixed at 0° and constant

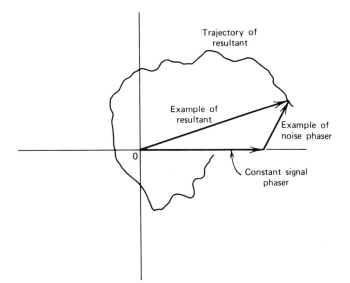

Figure 9.7   Phaser diagram of click generation.

amplitude while the noise adds to the signal with random, fluctuating amplitude and phase. The resultant traces out a continuous trajectory in the complex plane. A cycle is gained or lost—a click is generated—every time the trajectory encircles the origin.

A click is possible only if instantaneous noise exceeds the signal amplitude and if the phase of the noise goes through opposition with the phase of the signal. In the vicinity of threshold, the noise amplitude associated with a typical click event is likely to be just slightly stronger than the signal, so the click trajectory is likely to pass very close to the origin; that is, amplitude of the resultant often can be expected to be small in the middle of a click passage.

Under conditions of near-cancellation of signal by the noise, a small change of noise phase can cause a large change of resultant phase. Therefore, a click trajectory can sweep around the origin very rapidly, much more quickly than might be suggested by the restricted bandwidth of the input filter. These features of amplitude and phase have considerable bearing on the response of a PLD.

Clicks may also be examined by means of phase and frequency waveforms, as in Figure 9.8. In the absence of noise, the fixed-signal phaser of Figure 9.7 produces a constant resultant phase of 0°. Small noise causes small phase fluctuations about zero, while a click causes the resultant phase waveform to have a 2π step (Figure 9.8a).

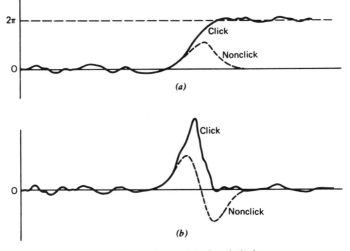

Figure 9.8   Click waveforms: (a) phase; (b) frequency.

Frequency is the time derivative of phase—the rate of rotation of the resultant phaser—and is sketched in Figure 9.8b. Small phase noise produces small frequency noise, while a phase step produces a large frequency spike; this is the spike or click that is heard in an audio message or observed in the laboratory.

Click waveforms vary widely; the only property they have in common is that each has an area of 2π or an integer multiple thereof. Polarity of a click pulse depends on whether a cycle is lost or gained. An individual pulse is essentially unipolar.

Clicks usually occur rapidly compared to the reciprocal of baseband signal bandwidth. To calculate the influence of clicks on output SNR_o, it is useful to approximate the waveform as an impulse with area 2π. An impulse has a flat spectrum extending down to DC and with substantial energy in the baseband.

Figure 9.8 also shows a large phase disturbance that does not cause a click: a nonclick. We see that the peak frequency output of the nonclick is much smaller than that of a completed click. More significantly, the frequency pulse associated with a nonclick is a doublet—which has its energy concentrated at high frequencies and falls off to zero at DC. A doublet causes much less disturbance to a lowpass system than does a unipolar pulse.

If average click rate is known, the contribution to output noise can be calculated.[6,8] The CNR at which clicks increase total noise one decibel above that calculated by (9.28) alone is the formal definition of threshold. There is substantial energy in a click, so threshold occurs at a surprisingly small click rate.

Rice[8] has determined the click rate for an ideal discriminator. It is a function of CNR, input passband shape, and modulation parameters. Use of his formulas provides a good prediction of threshold of an ideal discriminator.

Application to PLD

Unfortunately, no one has yet been able to analyze the output click rate of a phaselocked demodulator. The click concept gives some physical insight into the operation of a PLD, but a quantitative theory has been unattainable. This section summarizes the author's qualitative understanding of the problem, based on largely unpublished experimental work. (Smith[12,13] has pursued a PLD approach similar to that presented here, but neglecting the input filter.)

First, let us examine a complete block diagram of phaselocked FM demodulator, as in Figure 9.9. It is comprised of an input bandpass filter, a phase detector, a loop filter, a VCO, and lowpass post filter. All five elements are essential to proper operation of the PLD. Nonetheless, many earlier publications have ignored the input and output filters completely.

The post filter contains the deemphasis networks, correction for linear-filter distortion by the PLL transfer function, and the main baseband filtering of the recovered message. This filtering is needed to achieve the FM improvement of (9.30). However, the post filter processes the signal only after it is recovered from the PLL; it clearly can have no influence on tracking performance and therefore does not affect threshold. Neglect of the post filter is justified in a study of threshold phenomena.

On the other hand, neglect of the input filter is competely wrong. It is commonplace to think of a PLL as a narrowband device that combats noise by means of its narrow bandwidth. This is a misconception when

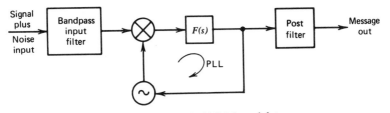

Figure 9.9 Phaselocked FM demodulator.

applied to a PLD where the loop bandwidth must be rather large. In fact, the loop bandwidth is likely to be considerably larger than the RF signal bandwidth.

The input filter must be wide enough to avoid excessive distortion of the message; this is a very complex subject in itself and is not treated here. A lower bound on input bandwidth is established by *Carson's rule*:

$$B_i \geqslant 2(\Delta f + f_m) \tag{9.31}$$

which is appropriate for sinusoidal modulation, or, a modified version

$$B_i \geqslant 2(B_m + \gamma \sigma_f) \tag{9.32}$$

which is appropriate for gaussian modulation with lowpass bandwidth B_m, rms frequency deviation σ_f, and "crest factor" γ (Chapter 4).

Experiments show that the best choice of loop bandwidth (discussed below) is substantially larger than the Carson's rule bandwidth. We can expect that the loop bandwidth will exceed the input-filter bandwidth in a well-designed loop so the loop provides no appreciable linear filtering of the RF noise. The only significant noise reduction is provided by the input filter. For this reason, the input filter should be as narrow as possible, consistent with signal-distortion specifications. A wider filter admits extra noise, thereby degrading performance.

This statement has been tested by experiment. The results show conclusively that excess bandwidth of the input filter increases the threshold of the PLD. The amount of degradation depends on the degree of bandwidth excess; some measured results are shown in Figure 9.10 for sinusoidal modulation and a first-order loop.

If frequency deviation goes out to the edges of the input filter passband, we must be concerned with the response in the edges. Most filters roll off gradually at frequencies away from center; frequency deviation of the signal imposes a corresponding amplitude variation on the filter output. This incidental AM is quite noticeable as a scalloping of the signal envelope applied to the PLL.

To a first approximation, the PLL output is insensitive to amplitude effects. However, when we look more closely, we recognize that the loop gain is proportional to signal amplitude so the scalloping produces an instantaneous reduction in gain. In consequence, the tracking ability of the PLL is impaired and the loop is less capable of tracking the modulation

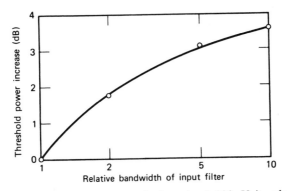

Figure 9.10 Threshold increase due to excessive input bandwidth. Unity relative bandwidth corresponds to (9.31). Ordinate shows additional signal power needed at threshold because of widened bandwidth.

excursions. Therefore, the scalloping worsens the threshold, particularly in a first-order loop where maximum loop stress occurs at maximum frequency deviation (see section on Modulation Limits in Chapter 4). A desirable bandpass filter would have a flat response over the entire frequency-deviation range.

If an input filter is good for use with a PLD it is presumably also good with a conventional discriminator. Using identical input filters, a PLD can exhibit a lower threshold than a conventional discriminator. How does the improvement come about?

Signal plus noise at the filter output will have click events as described above. The average rate of clicks is given by Rice's analysis. An ideal discriminator, by definition, demodulates every one of the clicks in the signal applied to it. A PLL is unable to follow some of the clicks, so its output remains closer to the original message than the output of an ideal discriminator. It is this inability to follow some of the input clicks that accounts for the improved threshold of a PLD.[14]

Why does the PLL fail to follow some clicks? One reason is that the PLL is a limited-bandwidth element and a typical click is quite fast; the loop often cannot move quickly enough to follow the click around the circle. The sluggishness is emphasized when the noise nearly cancels the signal so that amplitude of the resultant is very small as the trajectory whips around the origin. In the vicinity of threshold-CNR most click events are associated with small amplitudes of the resultant.

A small amplitude means reduced gain of the PLL and therefore reduced ability to follow the resultant phasor. Reduction of gain is a nonlinear effect that, in this instance, apparently improves the threshold behavior.

We are now in a position to see why a limiter would worsen threshold performance. An ideal hard limiter holds the amplitude of the resultant phasor constant, irrespective of any possible cancellation of signal by noise. If amplitude is large as the trajectory goes around the circle, the PLL is better able to follow and more input clicks are demodulated than when a limiter is omitted.

Hess[14] has shown the deleterious effects of a limiter by experiment and by approximate analysis. Minimizing threshold demands that a limiter be omitted. Therefore, any PD characteristic that implicitly uses a limiter—sawtooth, rectangular, triangular, or any sequential PD—should be avoided (Chapter 6).

Sometimes one encounters the notion that PLD threshold is somehow "caused" by the nonlinearity of the phase detector and that the threshold could be avoided—or at least reduced—if only a linear PD were possible. This is not the case. Instead, we argue here that the reduced threshold of a PLD is at least partly due to the nonlinearity of the PD and that a linear PD would yield the same threshold as an ideal discriminator.

Most real PDs have a periodic characteristic, whereas a linear PD would have some means of counting cycles and therefore has a straight-line characteristic extending to infinity in both directions. In equilibrium, both types of PD cause the loop to track close to the PD null.

What happens when a fast input click appears? Let us assume that the loop is too slow to follow the click immediately and just consider behavior after the click has ended. The periodic PD ignores the cyclic increment and the loop continues to track as if the click had never occurred. However, the linear PD recognizes that an extra cycle has been accumulated, so it produces an output corresponding to a phase error of 2π. The loop servos out that error by retarding the VCO phase by 2π in order to return to the PD null. In other words, the PLD with a linear PD is unable to ignore input clicks and eventually tracks them all, even if quite slowly. A linear PD is just as bad as an ideal discriminator. The periodic nonlinearity of a real PD contributes, in part, to the improved threshold performance of a PLD.

Clearly, to ignore as many input clicks as possible, the loop bandwidth should be as narrow as possible. If the bandwidth were very large the loop would follow all input clicks and performance would be the same as that of an ideal discriminator.

On the other hand, bandwidth must not be too narrow or else modulation will cause cycle slipping even in the absence of noise (Chapter 4). A loop that is overstressed by excessive modulation is very sensitive to slips induced by small noise disturbances.

At the demodulator output a cycle slip is indistinguishable from a demodulated input click; for convenience they are all called *output clicks*. It seems reasonable to suppose that a compromise bandwidth will minimize the output click rate.

Figures 9.11 through 9.13 show representative click-rate data gathered[15] on a second-order loop. Table 9.1 gives the experimental parameters. Click rate was measured by instrumentation described in Ref. 16. Each data point represents 100 sec of accumulating output clicks. (Data gathering is painfully slow.)

Figure 9.11 plainly bears out the prediction of an optimum loop bandwidth. Substantial improvement in click rate can be obtained by choosing the proper loop bandwidth. Or, stated negatively, a large penalty is incurred if the wrong bandwidth is used.

Figure 9.12 shows a broad click-rate minimum for damping in the vicinity of 1 to 2. Other data suggest that a choice of 1 to 1.5 ought to be good, in general.

Click-rate curves of Figure 9.13 are plotted versus CNR, with natural frequency as a parameter. The solid black curve labeled n_r is a plot of Rice's[8] prediction of click rate of an ideal discriminator with the same modulation parameters and the same input filter. Data points for a wideband PLL (40 kHz) agree very closely with the Rice prediction, confirming the statement that a wideband PLL has the same threshold as an ideal discriminator.

The solid straight line labeled $\bar{n}$ shows the click rate that will increase output noise of (9.28) by 1 dB. Threshold is formally defined by the intersection of $\bar{n}$ and an actual click-rate curve. Intersection for the optimum-bandwidth PLL is some 2.5 dB lower than that for the ideal discriminator.

Choice of optimum bandwidth is strongly dependent on modulation parameters; it is *a priori* knowledge of message statistics that permits threshold reduction. Design of a conventional discriminator virtually ignores message statistics, thereby incurring a threshold penalty. On the other hand, a conventional discriminator is relatively insensitive to changes of message statistics, whereas the PLD is affected adversely and perhaps catastrophically.

A second-order loop is not necessarily the configuration that provides minimum threshold. If modulation index is small (as in the examples shown), then a first-order loop has nearly the same performance (as long as any steady frequency offset is small compared to loop bandwidth K). If

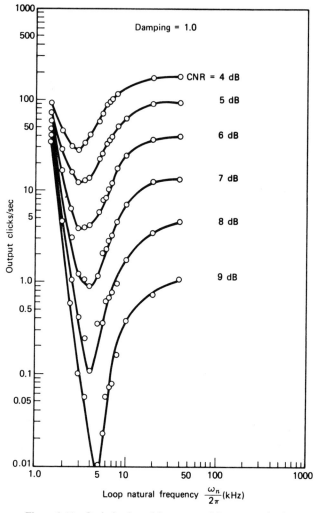

Figure 9.11 Optimization of loop natural frequency; $\zeta = 1$.

modulation index is large, then a higher-order loop tracks with less phase error (Figures 4.4 and 4.5). Experiments have shown second-order loops outperform first-order loops for large modulation index.

The experimental data shown here (Table 9.1 and Figures 9.11 to 9.13) were taken for a moderately small modulation index. Threshold extension of a PLD improves, relative to an ideal discriminator, as the modulation index increases.

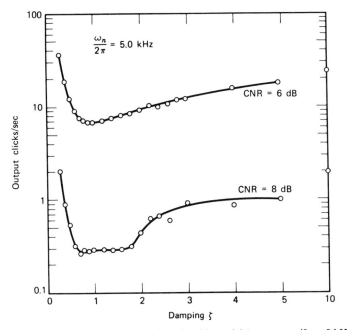

Figure 9.12 Optimization of loop damping. Natural frequency $\omega_n/2\pi = 5$ kHz.

Formal Optimization

A PLD is not the optimum FM demodulator but only an approximation thereto. The optimum demodulator would examine the entire message, even if it were of infinite duration, before producing the *maximum a posteriori* (MAP) estimate of the message. Viterbi[17] and Van Trees[18] give excellent discussions of MAP estimation applied to FM demodulation.

The integral equation of the MAP estimator is nearly identical to that of a phaselock loop; the only difference is in the limits of integration. On paper the difference seems trivial, but the PLL must track in real time—work with zero lag—and it cannot wait for the end of message before starting to process the signal.

Because of the close resemblance between integral equations, many investigators have hoped that the PLL would be a good approximation to the MAP demodulator. Better approximations can be obtained from complex digital schemes[19-22] but these are not PLLs.

Although we are barred from achieving the ultimate MAP performance, we can still ask, What is the optimum, zero-lag, stable PLD? To permit mathematical tractability, it is common to assume linear operation of the

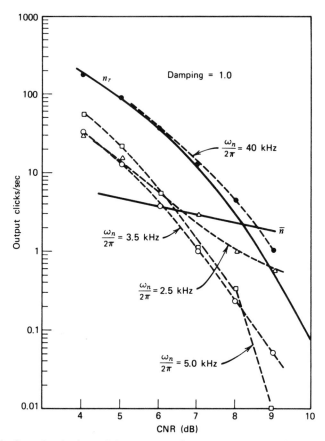

Figure 9.13 Second-order loop click rate. ($\triangle$)$\omega_n/2\pi = 2.5$ kHz; ($\bigcirc$) 3.5 kHz; ($\square$) 5.0 kHz; ($\bullet$) 40 kHz.

Table 9.1 Click-Rate Experimental Parameters

Modulation type: gaussian
Modulation spectrum: essentially flat from DC to 2.4 kHz ($B_m = 2.4$ kHz)
Deviation: $\sigma_f = 1485$ Hz
Spectral occupancy, equation 9.32: $2(B_m + \gamma\sigma_f) = 15.2$ kHz, for $\gamma = 3.5$
Input bandwidth: 15.2 kHz, -1 dB
                18.3 kHz, -3 dB
                24 kHz, -30 dB
                16.4 kHz, noise bandwidth
Noise bandwidth $2B_L$ of "optimum" PLL: 27.5 kHz ($\omega_n/2\pi = 3.5$ kHz; $\zeta = 1$)

PLL and to determine the optimum realizable* Wiener filter. Viterbi warns, emphatically, that this procedure does not lead to the MAP performance, nor does it lead necessarily to the optimum PLD. Nonlinear behavior cannot be inferred from linearized analysis.

The Wiener procedure makes use of signal and noise spectra in arriving at the optimum loop transfer function $H_o(\omega)$. If the frequency modulation is stationary, it can be represented by a modulation spectrum $\Phi_m(\omega)$ with units of $(\text{rad/sec})^2/\text{Hz}$. Formally, the spectrum of the resulting phase modulation of the transmitted signal is $\Phi_m(\omega)/\omega^2$ rad^2/Hz.

(A message need not be stationary. If it is not, the spectrum Φ_m does not exist and the Wiener procedure must be modified. Even if the frequency modulation is stationary, the phase modulation can be nonstationary, so the phase-modulation spectrum might not exist, rigorously speaking. However, the formal spectrum Φ_m/ω^2 serves our purposes adequately.)

Noise is represented by the equivalent baseband noise $n'(t)$, which has a spectral density $\Phi_{n'}(\omega)\,\text{rad}^2/\text{Hz}$ (see Chapter 3). Since the PLL is preceded by a bandpass filter with significantly narrow bandwidth, the density $\Phi_{n'}$ cannot even be approximated as white. It is important that the actual spectrum, as detailed in Chapter 3, be used in any analysis.

By using material from Chapters 3 and 4 and Appendix A, it can be shown that the linearized approximation to the phase error variance is

$$\sigma_e^2 = \frac{1}{2\pi} \int_0^\infty \left[|1 - H(j\omega)|^2 \frac{\Phi_m(\omega)}{\omega^2} + \Phi_{n'}(\omega)|H(j\omega)|^2 \right] d\omega \qquad (9.33)$$

The Wiener optimum filter $H_o(\omega)$ minimizes this phase variance.

Derivation of the optimum realizable Wiener filter is a tedious matter but is well covered in the literature.[23-26] If the two spectra can be represented as rational functions in ω^2 (almost any spectrum of engineering interest can be closely approximated in this manner), then the Wiener-optimized loop transfer function is

$$H_o(\omega) = \frac{1}{\left(\Phi_{n'} + \Phi_m/\omega^2\right)^+} \left[\frac{\Phi_m/\omega^2}{\left(\Phi_{n'} + \Phi_m/\omega^2\right)^-} \right]^+ \qquad (9.34)$$

where the superscripts $+$ and $-$ indicate that only the poles and zeros in the upper or lower halves, respectively, of the complex ω plane are to be taken.

---

*The unconstrained, optimum Wiener filter has infinite lag and is therefore unrealizable.

If either spectrum is of high order in ω^2 (as happens when the roll-off slope is steep), then $H_o(\omega)$ is also of a high-order transfer function and will be inconvenient to implement.

If $\Phi_m(0) = 0$, then the formally derived optimum loop filter has a zero at the origin, cancelling the integrator action of the VCO. We must not forget the lessons of Chapter 4; there always are phase and frequency offsets that can only be accommodated by unimpaired integration. The loop cannot phaselock if the integration is cancelled. Optimum loop design must take account of the offsets as well as the message spectrum.

The optimum Wiener filter minimizes loop-phase-error variance in the linear region of operation. However, our concern is in minimizing threshold; does the Wiener filter accomplish that also? No analytic method sheds any light on the question—we do not know how to analyze the PLL accurately in the threshold region. Only experiment is available to the investigator.

Few reports of experimental testing of Wiener-optimized loops can be found in the literature. By gleaning among several obscure sources, it was discovered that the linearly derived Wiener filter in fact *does not* minimize the threshold. At best, it is a starting point for an empirical search for the minimum-threshold PLD.

The potential complexity of a Wiener filter is often unacceptable. A simpler approach is to use a loop of ordinary form (for example, a standard second-order loop) and then minimize the threshold by adjusting the loop parameters (e.g., damping and natural frequency in the second-order loop).

An analytic approach could be attempted by explicitly writing $H(j\omega)$ in terms of its parameters in (9.33) and then minimizing by proper choice of parameters. If the spectra of modulation and noise are at all complex, the minimization must be accomplished by means of computer search methods.

Such a search was tried with a second-order loop and spectra as described in Table 9.1. The following results were obtained:

• Calculated values of phase-error variance were an oscillatory function of natural frequency, suggesting that any automatic computation procedures might get into difficulty.

• The minima were extremely shallow. This comes about because of the narrowband filter preceding the PLL; widening loop bandwidth beyond that of the filter bandwidth adds very little additional noise.

• Experiment did not agree at all well with the calculations. The natural frequency producing minimum click rate in the laboratory did not correspond with the natural frequency that yielded minimum calculated variance.

Before concluding that linearized phase variance is a poor way to approach PLD threshold optimization let us examine one other method. We represent the loop transfer function in polar form

$$H(j\omega) = |H|e^{j\psi} \tag{9.35}$$

where the frequency dependence of the polar components has been suppressed for notational convenience. Substituting (9.35) into (9.33) and performing some algebra, we obtain the variance as

$$\sigma_e^2 = \frac{1}{2\pi} \int_0^\infty \left[(1 - 2|H|\cos\psi + |H|^2) \frac{\Phi_m}{\omega^2} + \Phi_{n'}|H|^2 \right] d\omega \tag{9.36}$$

Irrespective of $|H|$, the variance will be minimized if $\cos\psi = +1$. (The same condition arises in the derivation of the optimum, Wiener, unrealizable, infinite-lag filter.[26]) That is, $\psi = 2\pi k$, where $k = \ldots -2, -1, 0, 1, 2, \ldots$.

Amplitude and phase response of a network are closely related; they cannot be specified separately.[27] Assuming a minimum-phase network and oversimplifying somewhat, the phase condition implies that the amplitude response has a slope of $24k$ dB/octave. From the discussion accompanying (9.22) we know that flat amplitude response is needed in the frequency range that includes the modulation spectrum, so a choice of $k = 0$ deserves attention.

In conventional PLLs that have equal numbers of poles and zeros in the loop filter to control damping and stability, the amplitude response is flat for low frequencies and falls off at -6 dB/octave at high frequencies. The phase is zero at DC and approaches $-90°$ asymptotically for high frequency.

A realizable network with zero phase at all frequencies must also have constant amplitude response at all frequencies; no finite-bandwidth PLL could have such response. Neglecting any parasitic elements, the rolloff of a conventional loop is due to the integrator action of the VCO; if that could be overcome, a phase near $0°$ could be achieved over a much larger frequency range than is accomplished in an ordinary loop.

By building a loop with proportional plus integral *plus derivative* control we achieve asymptotically zero phase at high, as well as low, frequencies. Such a loop is shown in Figure 9.14 (compare against Figure 2.2) and its Bode plots are shown in Figure 9.15.

Novick and Klapper[28,29] have arrived at essentially the same configuration, but starting from the phase-feedback circuit of Figure 6.14. They have devised a variance-minimization algorithm and have found minimum variance if $K\tau_1\tau_3/\tau_2 = 1$ (notation is shown in Figure 9.14). Therefore, at high frequencies $|G|$ approaches 1 and $|H|$ approaches 0.5. Phase of the

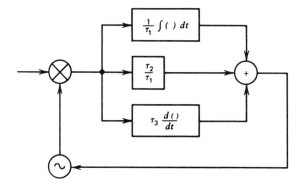

Open—loop gain:

$$G(s) = \frac{K_o K_d}{s} \left(\frac{\tau_2}{\tau_1} + \frac{1}{\tau_1 s} + \tau_3 s \right)$$

$$= \frac{K}{s^2 \tau_2} \left(s^2 \tau_1 \tau_3 + s \tau_2 + 1 \right)$$

$$\left(K = \frac{K_o K_d \tau_2}{\tau_1} \right)$$

Closed — loop gain:

$$H(s) = \frac{G(s)}{1 + G(s)} = \frac{s^2 \tau_1 \tau_3 + s \tau_2 + 1}{s^2 \left[\frac{\tau_2}{K} + \tau_1 \tau_3 \right] + s \tau_2 + 1}$$

Figure 9.14 PLL with derivative control added.

closed-loop approaches zero at high and low frequencies but must exhibit a lag in the vicinity of the break in open-loop amplitude slope. Zeros of G or H could be complex.

Since response is flat at high frequencies, the noise bandwidth of this circuit would be infinite. Yet the authors report a significant lowering of threshold from a conventional PLL. If the technique proves to be generally applicable it could be an important advance in the PLD art. More work is needed; in particular, a physical explanation of the cause of the improvement would be valuable.

The informed reader will object to the inclusion of an ideal differentiator, which is a nonphysical network. Any approximation must eventually flatten off from the +6 dB/octave response of a differentiator, which means that the ultimate high-frequency phase of the closed loop cannot be zero after all; it must be at least −90°. Note that this conclusion is based entirely on theoretical limits of realizable networks and does not rest on

the inevitability of additional rolloffs caused by unwanted parasitic elements. No information exists, at the present time, on the degree of
perfection required—on the permissible location of the unavoidable
pole(s)—of the practical differentiator.

Imperfections of real differentiators can be circumvented by using the
phase-feedback phase detector instead of a differentiator. The loop transfer functions will be similar to those shown in Figure 9.14.

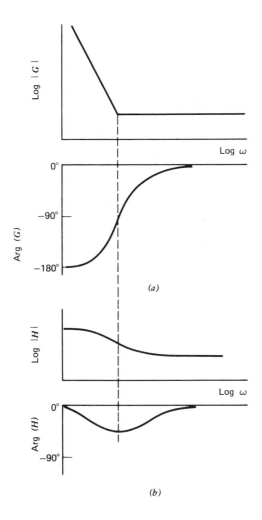

Figure 9.15 Bode plots of PLL with derivative control added: (a) open loop (compare
Figure 2.10), (b) closed loop.

PLD Threshold: A Summary

- An input filter is an integral part of a PLD and should not be neglected.
- Bandwidth of the input filter should be the minimum consistent with acceptable message distortion. A larger bandwidth entails a threshold penalty.
- Amplitude response of the input filter should be substantially flat over the full range of frequency deviation to avoid interaction with PLL gain.
- A limiter is not needed in a PLD and its inclusion raises threshold level.
- Bandwidth of the PLL must be substantially larger than that of the message and probably larger than that of the input filter.
- For any loop configuration, there exists an optimum loop bandwidth that yields a minimum output click rate. The optimum is a weak function of the input CNR and a strong function of modulation conditions. Present methods are inadequate to determine the optimum analytically; experiment with the actual signal and hardware to be used is perhaps the best approach open to the design engineer.
- If a second-order loop is employed, experiment suggests that damping of about 1 to 1.5 is optimum.
- A standard second-order loop is not likely to be the optimum filter configuration or even the best practical configuration.
- Wiener optimization of a linearized loop appears not to offer guidance to the practical minimum-threshold PLD.
- Adding derivative control seems to be helpful, but a physical explanation of its effect is lacking.
- Much more work remains to be done.

REFERENCES

1. H. S. Black, *Modulation Theory*, New York, Van Nostrand, 1953, Chap. 11.
2. W. B. Davenport and W. L. Root, *An Introduction to the Theory of Random Signals and Noise*, New York, McGraw-Hill, 1958, Chaps. 12 and 13.
3. S. Butman and V. Timor, "Interplex—An Efficient Multichannel PSK/PM Telemetry System," *IEEE Trans.,COM*-20, pp. 415–419, June 1972.
4. B. D. Martin, *The Pioneer IV Lunar Probe: A Minimum-Power FM/PM System Design*, Technical Report No. 32-215, Jet Propulsion Laboratory, Pasadena, CA, March 15, 1962.
5. W. C. Lindsey, "Design of Block-Coded Communication Systems," *IEEE Trans., COM*-15, pp. 525–534, August 1967.
6. J. Klapper and J. T. Frankle, *Phase-Locked and Frequency-Feedback Systems*, Academic Press, New York, 1972, Chap. 6.
7. M. G. Crosby, "Frequency Modulation Noise Characteristics," *Proc. IRE*, Vol. 25, pp. 472–514, April 1937.

8. S. O. Rice, "Noise in FM Receivers" *Time Series Analysis* (M. Rosenblatt, Ed.), Wiley, New York, 1963, Chap. 25.

9. O. Shimbo, "Threshold Characteristics of FM Signals Demodulated by an FM Discriminator," *IEEE Trans.*, *IT*-15, pp. 540–549, Sept. 1969. Corrections: *IT*-16, p. 769, November 1970.

10. O. Shimbo, "Threshold Noise Analysis of FM Signals for a General Baseband Signal Modulation and its Application to the Case of Sinusoidal Modulation," *IEEE Trans.*, *IT*-16, pp. 778–781, November 1970.

11. J. R. Carson, "Notes on the Theory of Modulation," *Proc. IRE*, Vol. 10, pp. 57–64, February 1922.

12. B. M. Smith, "Phase-Locked Loop Threshold," *Proc. IEEE*, Vol. 54, p. 810, May 1966.

13. B. M. Smith, "A Semi-Empirical Approach to the PLL Threshold," *IEEE Trans.*, *AES*-2, pp. 463–468, July 1966.

14. D. T. Hess, "Cycle-Slipping in a First-Order Phase-Locked Loop," *IEEE Trans.*, *COM*-16, pp. 255–260, April 1968.

15. J. F. Heck and F. M. Gardner, *Multiplex Communication System*, 1976 Final Technical Report, Lockheed Missiles and Space Co., D569571 Internal Report.

16. F. M. Gardner, "A Cycle-Slip Detector for Phase-Locked Demodulators," *IEEE Trans.*, *IM*-26, pp. 251–254, September 1977.

17. A. J. Viterbi, *Principles of Coherent Communications*, McGraw-Hill, New York, 1966, Chaps. 5 and 6.

18. H. L. Van Trees, *Detection, Estimation and Modulation Theory: Part II*, Wiley, New York, 1971, Chaps. 2–4.

19. P. K. S. Tam and J. B. Moore, "A Gaussian Sum Approach to Phase and Frequency Estimation," *IEEE Trans.*, *COM*-25. pp. 935–942, September 1977.

20. P. K. S. Tam and J. B. Moore, "Improved Demodulation of Sampled FM Signals in High Noise," *IEEE Trans.*, *COM*-25, pp. 1052–1054, September 1977.

21. D. W. Tufts and R. M. Rao, "Frequency Tracking by MAP Demodulation and by Linear Prediction Techniques," *Proc. IEEE*, Vol. 65, p. 1220, August 1977.

22. R. S. Bucy and A. J. Mallinckrodt, "An Optimal Phase Demodulator," *Stochastics*, Vol. 1, pp. 3–23, 1973.

23. Y. W. Lee, *Statistical Theory of Communication*, Wiley, New York, 1960, Chaps. 14–17.

24. W. B. Davenport and W. L. Root, *Random Signals and Noise*, McGraw-Hill, New York, 1958, Chap. 11.

25. D. Middleton, *Statistical Communication Theory*, McGraw-Hill, New York, 1960, Chap. 16.

26. H. W. Bode and C. E. Shannon, "A Simplified Derivation of Linear Least Square Smoothing and Prediction Theory," *Proc. IRE*, Vol. 38, pp. 417–425, April 1950.

27. H. W. Bode, *Network Analysis and Feedback Amplifier Design*, Van Nostrand, New York, 1945.

28. W. A. Novick and J. Klapper, "Optimum Design of the Extended-Range Phase-Locked Loop," *Conference Record of National Telecommunications Conference* 1972, Paper 32D.

29. W. A. Novick, "Investigation and Optimum Design of the Generalized Second-Order Phase-Locked Loop," PhD Dissertation, New Jersey Institute of Technology, Newark, 1976.

Chapter Ten

Locked Oscillators and Synthesizers

A phaselocked loop includes a locked oscillator, by definition. It may seem redundant to devote a separate chapter to locked oscillators in a book on PLLs. Nonetheless, there are applications in which the primary objective of the PLL is to lock an oscillator, usually to improve its stability. Some of those applications are presented here.

A frequency synthesizer generates a large number of different output frequencies, all related to a single, highly stable reference source. Phase-locked synthesizers are the most popular type. Some elements of the principles of phaselocked synthesizers are described in this chapter.

10.1 OSCILLATOR STABILIZATION

There are two diametrically opposed varieties of oscillator stabilization: narrowband and wideband. In the first, a narrowband PLL is employed as a filter to clean up another oscillator or other signal that is accompanied by noise. In the second, a noisy oscillator is phaselocked to a clean reference to stabilize the locked oscillator.

Crystal oscillators used as frequency standards have their best long-term frequency stability if they are operated at extremely low RF power levels (crystal aging is slower at the low levels). However, as is noted in Chapter 6, best short-term phase stability is obtained at an intermediate power level, where the RF signal is much greater than the circuit noise.

The best results are obtained if two separate oscillators are used: a very low-level one for good long-term stability and a second oscillator, phase-locked to the first, operated at a higher power level for good short-term stability. Bandwidth of the loop should be as narrow as possible, consistent with maintaining reliable lock, and output is taken from the locked oscillator.

Using the loop is equivalent to passing the phase noise of the first oscillator through an extremely narrow filter to reduce it substantially. The same technique is useful for cleaning up the output of frequency synthesizers in which harmonics and multiplier products are often present.

Microwave oscillators with usefully large power output can be built with transistors, klystrons, backward-wave tubes, IMPATT or TRAPATT diodes, or Gunn diodes. Electronic tuning is accomplished by changing operating biases on the active device, by using a varactor diode, or by using a magnetically variable *YIG* resonator. These diverse oscillators share a common trait of poor phase stability. Without additional stabilization, they are unusable in narrowband applications.

An effective method of stabilization is to lock the microwave oscillator to a harmonic of a stable, low-frequency source, such as a crystal oscillator. The loop tracks out the phase fluctuations of the locked oscillator (see Chapter 6), so the output has the stability of the frequency-multiplied reference source.

One configuration of oscillator stabilization is shown in Figure 10.1. The only novel element is the frequency multiplier needed to obtain the proper harmonic of the stable low-frequency source. Often the multiplier is incorporated into the phase detector itself (see next section). Large numbers of such phaselocked oscillators are sold as complete packages with acquisition circuitry included. The packages are widely used for transmitters and for receiver local oscillators in fixed-frequency service. A phaselocked source is usually more economical of power than a multiplier string of equal output power.

In a laboratory instrument, it is necessary to be able to operate at many different frequencies, ideally over a large, continuous band. This is accomplished by the heterodyne stabilizer or *lockbox* of Figure 10.2. A sample of microwave output is heterodyned against the Nth harmonic of a stable reference oscillator at frequency f_r to generate a mixer product at the intermediate frequency. This is filtered, amplified, perhaps limited or otherwise level controlled, and applied to a phase detector where it is compared in a phase against a low-frequency oscillator at frequency f_x.

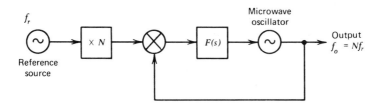

Figure 10.1 Locked microwave oscillator.

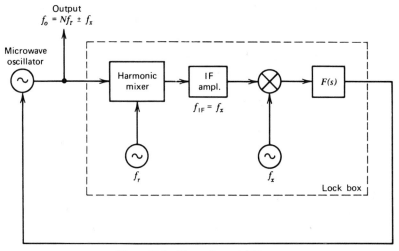

Figure 10.2 Heterodyne stabilizer.

The DC output of the PD is filtered and used to phaselock the microwave oscillator.

Harmonic mixers are wideband devices and produce a great many harmonics of the reference oscillator. (A harmonic comb generator might be used to enhance the operation of the mixer alone.) The circuit is designed to operate at any of the harmonics in a specified range of frequencies; the loop locks to any one of those harmonics, which are spaced f_r apart. Finer coverage is obtained by varying f_r or even f_x.

A harmonic mixer is typically inefficient, so the desired IF output is weak as compared to the two RF input levels. Low amplitude is easily remedied by an ordinary IF amplifier, provided that the IF signal amplitude does not drop too close to the noise level. Since IF amplifiers are simpler to build and less costly than efficient, high-order harmonic generators, this is an effective method of harmonic locking. Furthermore, it is easier to build phase detectors at a reasonable intermediate frequency than in the microwave region.

Total phase noise of the locked oscillator is a filtered combination of the jitter of the unlocked microwave oscillator, the reference oscillator, and circuit noise. (Noise contributions of the f_x oscillator often can be neglected.) Oscillator noise is analyzed in Chapter 6 and additive noise is analyzed in Chapter 3.

We designate the phase noise spectrum of the unlocked microwave oscillator as $\Phi_o(\omega)$ rad^2/Hz, the phase noise spectrum of the reference oscillator as $\Phi_r(\omega)$, and the spectrum of the equivalent additive noise as

$\Phi_{n'}(\omega)$. If the IF noise may be considered white with spectral density N_o V^2/Hz, then $\Phi_{n'} = 2N_o/V_s^2$ rad^2/Hz, where V_s is the peak amplitude of the IF signal. Oscillator jitter spectra usually must be obtained experimentally or from manufacturer's specification.

The phase noise spectrum of the locked oscillator becomes

$$\Phi_p(\omega) = (\Phi_{n'} + N^2\Phi_r)|H(j\omega)|^2 + |1 - H(j\omega)|^2\Phi_o \qquad \text{rad}^2/\text{Hz} \quad (10.1)$$

The factor N^2 appears because $\times N$ multiplication magnifies phase excursions of the reference oscillator by N times.

Given the three noise spectra, the design task is to choose the transfer function to minimize output noise. Practically, the loop is almost always of second order, so the designer must select damping and natural frequency to minimize jitter.

The performance criterion is not uniquely specified. Sometimes one tries to minimize total jitter over a large bandwidth, but other applications place greater weight on minimizing the spectral density at a spot frequency. If the three constituent spectra are known analytically, then (10.1) or its integral can be minimized on ω_n and ζ. More likely, the spectra will not be known, so experimental adjustment of loop parameters will be needed to find optimum performance.

If the microwave oscillator is particularly noisy or if the reference is especially stable, the optimum bandwidth might be quite large. Typical units have bandwidths $K/2\pi \approx 20$ kHz on up to 1 MHz or so. As is noted in Chapter 8, an actual feedback loop has many more roll-off elements than are wanted. It is necessary to assure stability of the loop, and a satisfactorily stable design could be appreciably narrower than the ideal optimum derived solely from the noise spectra.

Along with destabilizing the loop, extra poles and delays are likely to induce peaking of the loop response, causing minor lobes to appear on an otherwise unimodal oscillator noise spectrum.

10.2 HARMONIC LOCKING

An oscillator often must be locked to a harmonic of the input reference frequency. The preceding section illustrates methods employing frequency multipliers to generate the harmonic in a straightforward manner. Another popular technique (Figure 10.3) utilizes frequency dividers to reduce the oscillator frequency to that of the reference. The technique is especially attractive at frequencies low enough to permit the use of digital counters as the dividers.

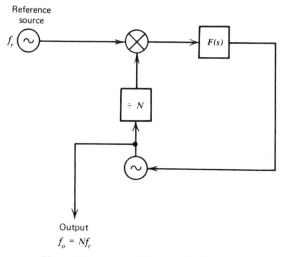

Output
$f_o = Nf_r$

Figure 10.3 Harmonic locking by division.

As in any harmonic generator, the phase jitter at the output includes a component equal to N times that portion of the reference jitter that passes through the loop transfer function. Also, if there is closed-loop baseband noise v_n at the PD output, then the corresponding VCO phase jitter is Nv_n/K_d, assuming that the spectrum of v_n lies inside the loop bandwidth. If N is large, the output jitter can be unacceptable, even for respectably small values of reference jitter or v_n. Extreme measures are sometimes needed to suppress stray circuit noises that are ordinarily negligible.

Harmonic locking can also be obtained without any distinct frequency multipliers or dividers in the circuit; the harmonics are produced by the phase detector itself. In a multiplier-type phase detector (Chapter 6) the useful DC output is the average product of the two inputs; DC output is possible only if both inputs have a component at the same frequency. It is not necessary that both inputs have the same fundamental frequency— only that they contain a common harmonic. Moreover, it is not essential that the actual input waveform contain the harmonic; it could also be generated by nonlinearity of the phase detector.

In consequence of this property of multipliers, a phase detector is capable, in general, of generating locking DC outputs for frequencies $Nf_r = Mf_o$, where M and N are integers. In the previous circuits M has been unity and only *integer harmonic locks* have been obtained. It is entirely possible to have $M \neq 1$ and thereby obtain *fractional harmonic locks*. The integer harmonic locks are best known; if either PD input is sinusoidal, only integer-harmonic locks are possible.

As an example, we consider a simple full-wave switching phase detector (Figure 6.5) in which $M=3$ and $N=1$. (Locking is to a subharmonic of the reference, but harmonic locking could be obtained by switching the PD from the reference so that $N=3$ and $M=1$.) Waveforms are shown in Figure 10.4.

Close inspection of the waveforms reveals the following facts regarding the PD output:

- Areas of positive and negative portions of the output waveform are unequal, which means that a DC component is present. Relative phase shift between the two inputs would alter the amount of DC—in this case, sinusoidally. The DC versus phase characteristic is repetitive in the period of the common harmonic.
- Only a comparatively small portion of the output waveform contributes to the useful DC (compare Figure 6.5) so the maximum possible DC amplitude is small. In general, high-order harmonics produce small locking voltages.
- The lowest beat-frequency component is at the difference frequency between the two inputs $f_r - f_i$. The beat-note can be much more prominent than the locking DC.

These results are typical of all harmonic PDs and are not specific to the example.

A conventional switching PD, driven with symmetrical square waves on both inputs, works only for odd values of M and N; a symmetrical square wave contains only odd harmonics. To operate on even integers, the waveforms must be asymmetrical and must contain appropriate even harmonics. A sampling PD works on any harmonic, since an impulse sampling train contains all harmonics of the sampling rate.

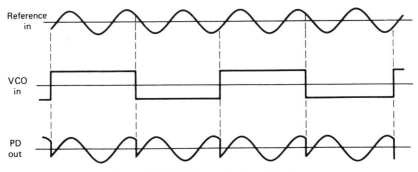

Figure 10.4 1/3-harmonic phase detection.

Conventional PDs are found in locked oscillators as in Figure 10.1, with multiplication of 3, 5, or even 7 times accomplished in the phase detector. If higher ratios are used, not only does the locking voltage become small, but there could be a problem in selecting the correct harmonic.

Fractional-harmonic locking is more likely to appear as an unwanted effect than to be attempted deliberately. Spurious locking at ratios as high as 17 : 18 have been observed experimentally. Making one input sinusoidal prevents all fractional-harmonic locks. If sinusoids are not feasible (many applications using digital components require square waveforms), then frequency-aiding acquisition circuits (Chapter 5) can often overcome the relatively weak spurious lock.

Despite the fact that they are usually detrimental, fractional-harmonic locks do offer some promise of a low-cost method of producing fractional harmonics. It might not be a bad solution for a specialized, fixed-frequency synthesizer.

The above discussion is centered on multiplier-type phase detectors, but simple flip-flop sequential PDs can also be shown to produce harmonic lock voltages—both integer and fractional. A general explanation of the multiplier action was easily given, but it is more difficult to generalize about sequential operations. However, tracing the waveforms of any specific sequential PD will quickly show its harmonic-lock properties.

Note that the sequential phase-frequency PD (Chapter 6) ought not support harmonic locking because the built-in frequency slewing should be capable of overcoming any harmonic lock voltages that could be generated.

10.3 TRANSLATION LOOPS

A frequency translator shifts an input frequency f_1 to an output frequency $f_1 \pm f_b$. The benefits of translating by means of a PLL may be seen from an example. Let us suppose we wished to offset a 30-MHz signal by 1 kHz. One way to accomplish this would be by means of conventional single-sideband techniques, but good suppression of carrier and rejected sideband would depend on critical circuit adjustments.

A phaselock offset could be completely noncritical if obtained as shown in Figure 10.5. In this technique a VCO, whose uncontrolled frequency f_o is close to the desired output, is heterodyned with the incoming frequency f_1; the beat-note is close to the desired offset f_b. This beat is compared with an oscillator whose frequency is exactly f_b, and the loop is closed back to the VCO so that the mixer beat-note is locked to the offset oscillator.

At first appearance, it would seem that phaselock has completely eliminated the residual carrier and unwanted sidebands that remain in

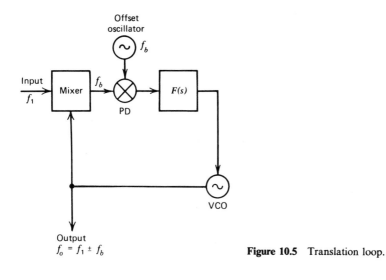

Figure 10.5 Translation loop.

conventional SSB techniques. Such perfection is not actually obtainable; any phase-detector ripple will modulate the VCO and produce unwanted sidebands in the output. If a full-wave phase detector is used, the input carrier, in principle, will not appear, and the dominant sidebands will be at $f_1 \mp f_b$ and $f_1 \pm 3f_b$. If a half-wave phase detector is used, the first-order sidebands will be at f_1 and $f_1 \pm 2f_b$; the undesired sideband at $f_1 \mp f_b$ is second order.

Ripple may be reduced to any desired extent by means of brute-force, noncritical lowpass filtering in the loop filter. Such filtering usually requires a narrowing of loop bandwidth. In essence, the unwanted sidebands are filtered by the PLL itself.

Notch filters are sometimes used; they combine large attenuation, at a fixed frequency, with small phase shifts at low frequencies within the loop bandwidth. Notch filters, when applicable, can be more effective than simple filters and may permit a larger loop bandwidth.

A phase detector with low inherent ripple (e.g., sample and hold PD or sequential phase-frequency PD; see Chapter 6) may be more effective than any practical filtering in suppressing ripple.

The loop of Figure 10.5 is capable of locking to either of the two sidebands $f_1 + f_b$ or $f_1 - f_b$. In most instances a specific sideband is wanted and the other is an undesired image. Provision must be made to avoid locking to the image.

If the VCO is unable to tune to the image, the problem vanishes. More generally, if the VCO can tune both to the desired signal and to the image, then other measures must be taken to avoid image lock.

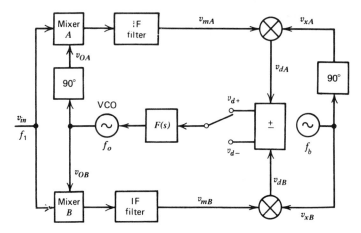

Figure 10.6 Image-rejecting PLL. (Waveform equations in Table 10.1.)

One approach is to use an image-rejecting phase detector as shown in Figure 10.6. This circuit is a variation on conventional single-sideband mixers and is applicable in any long loop where image rejection is needed.

The input signal at frequency f_1 is heterodyned to the offset or IF frequency in a pair of identical mixers. Filters select the desired difference frequency and reject the unwanted mixer products: in particular, the sum frequency. Because the two mixers have quadrature local drive, the two IF outputs are 90° apart in phase. (The same 90° relation could be achieved with a single mixer and a 90° phase-splitting network at the intermediate frequency. However, it is often convenient to produce quadrature drives using binary counters as in Figure 10.7.)

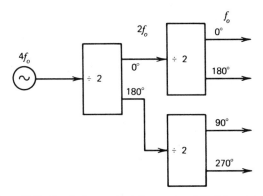

Figure 10.7 Quadrature outputs by means of digital counters.

Table 10.1   Waveform Equations* of Image-Rejecting PLL

$$v_{in} = \sin(\omega_1 t + \theta_1)$$

$$v_{oA} = 2\sin(\omega_o t + \theta_o)$$
$$v_{oB} = 2\cos(\omega_o t + \theta_o)$$

$$v_{mA} = \cos[(\omega_1 - \omega_o)t + \theta_1 - \theta_o]$$
$$v_{mB} = \sin[(\omega_1 - \omega_o)t + \theta_1 - \theta_o]$$

$$v_{xA} = 2\sin(\omega_b t + \theta_b)$$
$$v_{xB} = 2\cos(\omega_b t + \theta_b)$$

$$v_{dA} = \sin[(\omega_1 - \omega_o + \omega_b)t + \theta_1 - \theta_o + \theta_b]$$
$$\quad - \sin[(\omega_1 - \omega_o - \omega_b)t + \theta_1 - \theta_o - \theta_b]$$
$$v_{dB} = \sin[(\omega_1 - \omega_o + \omega_b)t + \theta_1 - \theta_o + \theta_b]$$
$$\quad + \sin[(\omega_1 - \omega_o - \omega_b)t + \theta_1 - \theta_o - \theta_b]$$

$$v_{d+} = (v_{dB} + v_{dA}) = 2\sin[(\omega_1 - \omega_o + \omega_b)t + \theta_1 - \theta_o + \theta_b]$$
$$\quad = 2\sin(\theta_1 - \theta_o + \theta_b) \text{ at lock. Selects } f_o = f_1 + f_b.$$
$$v_{d-} = (v_{dB} - v_{dA}) = 2\sin[(\omega_1 - \omega_o - \omega_b)t + \theta_1 - \theta_o - \theta_b]$$
$$\quad = 2\sin(\theta_1 - \theta_o - \theta_b) \text{ at lock. Selects } f_o = f_1 - f_b.$$

*Notation is defined in Figure 10.6.

Each IF signal drives a phase detector whose other input comes from the offset oscillator at frequency f_b. A 90° phasing is imposed between the f_b drive voltages to the two phase detectors.

Individual phase detectors have outputs at sum and difference frequencies, as shown in the waveform equations of Table 10.1. At lock either the sum frequency or the difference frequency of the PD goes to zero. In an ordinary loop, lock occurs at either zero-beat condition. In the image-rejecting PLL, adding or subtracting the two PD outputs cancels out either the difference- or sum-frequency component, respectively. Lock can occur only at the remaining component. High-side or low-side translation is determined by selecting sum or difference of the PD outputs.

Perfect cancellation is impossible to achieve, so a weak lock may still be possible at the image frequency. Acquisition aiding, such as rapid sweep, is needed to override locking to the weak image.

If the phase detectors are perfect multipliers and if the two inputs to each PD are sinusoidal, then the image-rejecting connection suppresses the double-frequency ripple.[1,2] This feature permits loop bandwidth to exceed the input frequency f_1—an impossible condition in a conventional loop.

As an interesting curiosity, if the input can be furnished as a two-phase signal, the image-rejecting PLL can track even if the input frequency f_1 goes through zero. One mechanization is shown in Figure 10.8.

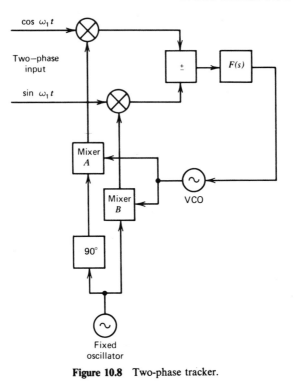

Figure 10.8 Two-phase tracker.

A complete image-rejecting PLL is often overly elaborate for merely avoiding lock to the unwanted sideband. A simple quadrature phase detector (Figure 5.15) has one polarity of output for lock on the upper sideband and the opposite polarity on the lower sideband. This polarity information can be furnished to acquisition-aiding circuits to prevent the loop from locking up at the wrong sideband.

10.4 FREQUENCY SYNTHESIZERS

A block diagram of a basic phaselocked synthesizer is shown in Figure 10.9. The synthesizer contains a reference source at frequency f_r and a VCO at frequency f_o. The reference frequency is divided by an integer N and the VCO frequency is divided by M; the two divided waves are then compared in a phase detector. Phaselocking imposes the condition of $f_r/N = f_o/M$, so the output frequency is locked to a rational fraction of the reference.

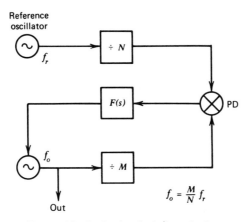

Figure 10.9 Basic phaselocked synthesizer.

Long-term stability and accuracy of the output frequency therefore are the same as that of the reference multiplied by M/N. Short-term stability is that of the reference times M/N if loop bandwidth is large and it is that of the VCO if loop bandwidth is small. The phaselocked synthesizer offers a means of generating a large number of highly accurate output frequencies at low cost.

Frequency selection is performed by changing the divider ratios M and N. Digital counters are used almost exclusively in this service; many clever counting schemes have been devised.[3]

Output frequency of the basic synthesizer is selectable in increments of f_r/N, which is the frequency at which phase comparison is performed. Loop bandwidth must be substantially smaller (assuming a conventional PLL and not an image-rejecting circuit) to suppress ripple adequately and to assure loop stability. If the desired increments are small, then the loop bandwidth must be extremely small.

On the other hand, a large loop bandwidth is preferred so as to achieve rapid acquisition and to stabilize short-term jitter of the VCO. A severe conflict can arise between these competing goals.

One method that sometimes alleviates the conflict is shown in Figure 10.10. The VCO frequency is P times the output frequency of $f_r M/NP$. Frequency increments of f_r/NP are obtained, but the phase comparison is performed at a frequency of f_r/N. The bandwidth conflict is relieved by a factor of P at the cost of operating the VCO dividers at P times the frequency required in the basic configuration. The technique is an economical solution to a serious problem, but limitations on counter speed inhibit its general application.

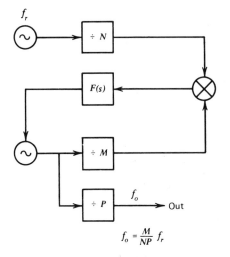

$$f_o = \frac{M}{NP} f_r$$

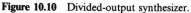

Figure 10.10   Divided-output synthesizer.

The multiple loops of Figure 10.11 combine harmonic loops, output division, and translation loops to avoid the conflicts of the basic loop. As shown, the example uses the same comparison frequency f_r/N at each phase detector, but this is neither necessary nor particularly desirable.

If we assume that all mixers select the difference product and that $f_1 > f_2/P_2$ and $f_2 > f_3/P_3$, then the output frequency is

$$f_1 = \frac{f_r}{N}\left[M_1 + \frac{M_2}{P_2} + \frac{M_3}{P_2 P_3} \right] \tag{10.2}$$

The output frequency is selectable in increments of $(f_r/N) \times 1/P_2 P_3$. More loops can be added to achieve even smaller increments without reducing the comparison frequency. The lower loops might all be constructed as identical modules for ease of manufacturing.

Mixers produce many unwanted output products as well as the one desired component. Existence of unwanted products raises the possibility of spurious components in the synthesizer output or even locking to the wrong frequency. Although not obvious from the diagram, presence of mixers also reduces the possible range of output frequencies.

A major design problem in synthesizers is to select internal frequencies so that mixer products do not cause adverse effects. That task is eased by the arrangement of Figure 10.11 in that each mixer is contained within and is filtered by its own loop. The mixer from one loop does not insert its output—either desired component or undesired component—directly into any other loop. The associated loop acts as a tracking filter for the desired

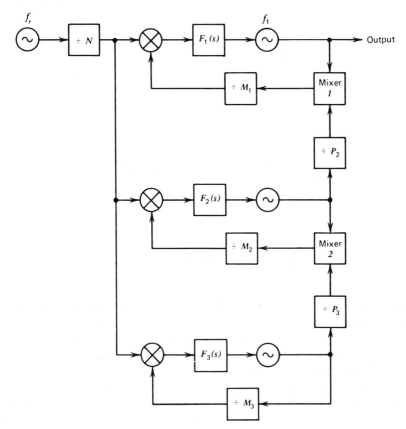

Figure 10.11 Multiple-loop synthesizer.

mixer product and suppresses all others in an amount depending on the loop bandwidth and transfer function. Ripple must be taken into account as in Section 10.3.

Another possible arrangement, shown in Figure 10.12, in essence, provides a vernier effect. The output frequency can be shown to be

$$f_1 = \frac{f_r}{(N+1)} \left[(M_1 + M_2) + \frac{M_1}{N} \right] \qquad (10.3)$$

If $M_1 + M_2$ is maintained constant, then changing M_1 by one will change the output frequency by an increment of $f_r/N(N+1)$. Additional loops can be stacked into the system to obtain even smaller increments. The M-divider settings are not independent as in (10.2); rapid computation of the

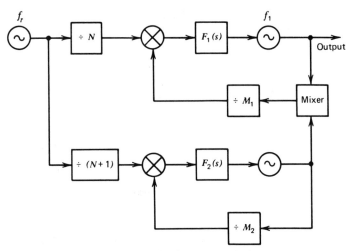

Figure 10.12   Vernier loop synthesizer.

correct M values for a specific frequency may be troublesome in a multiloop synthesizer.

All the preceding counters operate on an integer subharmonic of their input frequency. There is one pulse out for each M input pulses.

It is also possible to arrange counter circuits such that there are L pulses out for M input pulses[4]; the circuit multiplies by L/M. If the input frequency is f_i, then the counter output frequency is Lf_i/M, which is the comparison frequency for the PLL. However, by changing L, frequency increments of f_i/M can be obtained.

The L/M counter is a base-M counter that counts by increments of L instead of 1. Bilevel output depends on whether the count is greater than or less than $N/2$. The counter operates cyclically modulo M rather than resetting upon overflow.

Unfortunately, the L output pulses are not evenly spaced and cannot be if L and M are relatively prime. In effect, there is inherent phase modulation of the output pulse train. Spectral analysis[4] shows that all harmonics of f_i/M are present at the counter output, with Lf_i/M being the strongest. If M is large, the unwanted harmonics might be suppressed to acceptably small levels.

A synthesizer is usually a complex machine and many problems arise in its design. Spectral purity is always a goal and often demands extraordinary efforts to achieve. Oscillator noise is described in Chapter 6, the magnifying effects of harmonic locking are discussed in Section 10.1, and external additive noise is described in Chapter 3. Noise also arises within loop components such as op amps, phase detectors, and dividers. Digital circuits produce switching noise. Power supplies contribute unwanted hum. Mixers generate unwanted sidebands. Phase detectors have ripple in their outputs (Chapter 6).

These problems can be attacked in various ways. Out-of-band external disturbances are suppressed by the filtering action of the PLLs themselves; this is one of the major features of phaselocked synthesizers. In-band disturbances must be treated by suppressing the disturbance at its source —a task that, demands painstaking care to find and reduce all noise sources and skill on the part of the designer to devise circuits that introduce little disturbance (for example, phase detectors with small ripple; synthesizer configurations with modest divider ratios to minimize phase-jitter magnification). Good out-of-band filtering must be supported by excellent shielding and isolation of circuits operating at different frequencies.

Another problem is achieving fast acquisition of the correct lock frequency in each loop. Speed of acquisition depends on loop bandwidth —which is a strong reason to utilize a large comparison frequency. Acquisition aids (Chapter 5) help obtain fast, reliable lock. Since the intended frequency is known (in contrast to locking to an incoming signal of unknown frequency), a synthesizer often applies an analog pretuning voltage to the VCO to set the frequency nearly correct even before lock is obtained. This expedient reduces the frequency-search interval, thereby improving acquisition speed and avoiding possible frequencies of improper lock. (False locks and sideband locks are discussed in Chapter 8 and harmonic locks are discussed in Section 10.2. Improper locking can also occur if a signal from one circuit leaks through to another circuit because of inadequate shielding or isolation.)

Synthesizer design cannot be given justice in a small space; several books[5-7] provide vastly more detail plus a large bibliography. A good survey of synthesizer types and problems is given in Ref. 8. Perusal of the references will show that the synthesizer types presented here constitute only one of many types that have been used, even among the restricted class of phaselocked synthesizers. At the time of this writing, the phase-locked synthesizer with digital counters is by far the most popular of all.

Because the circuit uses digital counters, probably a "digital" phase detector (phase-frequency detector of Chapter 6), and perhaps even digital mixers[3]*, the synthesizers shown in Figures 10.9 to 10.12 are often called "digital" synthesizers. The designation is a misnomer; these are really analog loops. See Section 6.6 for a discussion of the requisite features of a

---

*Digital mixers—really D flip flops—encounter timing difficulties and are not recommended in high-performance applications.

true digital PLL. A digital synthesizer generates a sequence of digital numbers (which may be converted to analog form) by means of a difference equation[9, 10] or a table lookup[11], or by other techniques not yet known. It does not have a continuously running VCO for its output nor does it have an analog quantity as its error measurement, both of which are found in the counter phaselocked synthesizers.

REFERENCES

1. G. L. Baldwin and W. G. Howard, "A Wideband Phaselocked Loop Using Harmonic Cancellation," *Proc. IEEE*, Vol. 57, p. 1464, August 1969.
2. R. E. Scott and C. A. Halijak, "The SCEM-Phase-Lock Loop and Ideal FM Discrimination," *IEEE Trans.*, COM-25, p. 390, March 1977.
3. *Phase-Locked Loop Data Book*, 2nd ed., Motorola, Inc., Phoenix, AZ, August 1973.
4. D. G. Messerschmitt, "A New PLL Frequency Synthesis Structure," *IEEE Trans.*, COM-26, pp. 1195–1200, August 1978.
5. V. F. Kroupa, *Frequency Synthesis*, Wiley, New York, 1973.
6. V. Manassewitsch, *Frequency Synthesizers*, Wiley, New York, 1976.
7. G. Gorski-Popiel (Ed.), *Frequency Synthesis: Applications and Techniques*, IEEE Press, New York, 1975.
8. J. Noordanus, "Frequency Synthesizers—A Survey of Techniques," *IEEE Trans.*, COM-17, pp. 257–270, April 1969.
9. B. R. Saltzberg, "Frequency Modulation using Digital Filtering Techniques," *IEEE Trans.*, COM-18, pp. 632–637, October 1970.
10. K. Furuno, S. K. Mitra, K. Hirano, and Y. Ito, "Design of Digital Sinusoidal Oscillators with Absolute Periodicity," *IEEE Trans.*, AES-11, pp. 1286–1299, November 1975.
11. J. Tierny, C. M. Rader, and B. Gold, "A Digital Frequency Synthesizer," *IEEE Trans.*, AU-19, pp. 48–57, March 1971.

Chapter Eleven

Data Synchronizers

To an ever-increasing extent, information is being transmitted in digital formats. Very often, the data are transmitted *synchronously*: that is, in continuous, uniform pulse streams. Optimum detection of the data requires a local clock generator that is in close phase agreement with the received pulse train.

A data stream often must be modulated onto a carrier frequency before it can be transmitted over the communications channel. The most-efficient modulation methods are *coherent*; they make use of the phase information of the carrier. Optimum demodulation requires a local carrier source at the receiver whose phase is in close agreement with that of the received signal.

The receiver circuits that generate the receiver carrier and clock waveforms are known as the carrier and clock *synchronizers*. Their location within a typical receiver is shown in Figure 11.1. Phaselock loops are widely used in these synchronizers.

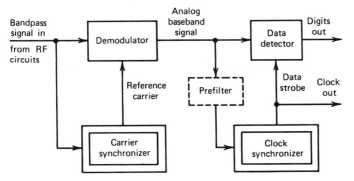

Figure 11.1 Digital receiver illustrating locations of synchronizers.

11.1 GENERAL PRINCIPLES

Efficient modulation techniques suppress the carrier completely; all transmitted energy resides in the data sidebands and none is "wasted" on a
discrete carrier component. (If a reliable carrier component is present, then
the carrier-tracking loops of earlier chapters can be used and the
suppressed-carrier methods discussed here are superfluous.)

Efficient data pulse streams contain no discrete component at the clock
frequency. For example, standard, square-waveform, random, non-return-
to-zero (NRZ) signaling actually has a spectral null at the clock frequency.

All the narrowband PLLs considered so far require a discrete signal
component at the frequency to be tracked. Since such a component is
absent from efficient data signals, a conventional PLL will fail to track and
is incapable of acting as a data synchronizer.

Suitable nonlinear circuits can regenerate a carrier or clock, respectively,
from a data signal that contains neither. Therefore, a nonlinear *regenerator*
is an integral portion of a data synchronizer; this particular nonlinearity
has not appeared in the PLLs studied up to now.

Two different general configurations of regenerator are shown in Figure
11.2. In one approach (Figure 11.2*a*) the regenerator is an entirely separate

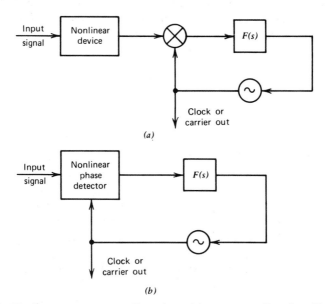

Figure 11.2 Nonlinear regenerator configurations: (*a*) separate nonlinearity; (*b*) nonlinear
phase detector.

circuit; it regenerates the desired carrier or clock, which is then tracked in the normal manner by an ordinary PLL.

In the other method, the phase detector itself contains additional nonlinearities that produce a useful error voltage from a data signal lacking carrier or clock component. Figure 11.2b shows the general configuration, and specific examples are given below.

(The separate nonlinearity does not require a PLL; a narrow bandpass filter could be used instead. Filters have been used to avoid the hangup of PLLs[1] when fast acquisition is needed (see Chapter 5) and also are found in large numbers in the telephone network.[2] A PLL can provide automatic compensation for frequency drift of either the signal or circuit components —a benefit not easily obtained with a simple filter circuit.)

The following steps must be taken in the design of a synchronizer:

• Devise a suitable nonlinearity for the specific task at hand.
• Identify disturbances and analyze performance.
• Select PLL parameters.

Numerous examples of nonlinear circuits are given in the following pages. Although the list could hardly be complete, there is enough to provide guidance to a designer.

The second step is the most difficult. Many disturbances are identified here, but the reader needing analysis details is referred to the literature. The essential nonlinearity usually makes analysis very difficult.

Loop parameters are selected much as for a conventional loop as is described earlier in the book. Once performance analysis has been accomplished, loop-parameter selection is comparatively simple.

11.2 CARRIER SYNCHRONIZERS

To introduce carrier synchronizers, it is useful to restrict our initial attention to binary signaling with rectangular NRZ bit waveforms: binary phase shift keying (BPSK). More general formats are considered afterward. If the data bit is a 1, the bandpass signal is transmitted with a phase of $+90°$; if the data bit is a zero, the signal phase is $-90°$. Each pulse is exactly T secs in duration and the pulses are T secs apart. If equal numbers of 1s and 0s are transmitted, the carrier is completely suppressed. If the baseband data stream waveform is represented as $m(t)$, the transmitted signal is

$$v_s(t) = m(t)\sin(\omega_i t + \theta_i) \tag{11.1}$$

where ω_i is the carrier frequency and θ_i is the signal phase, as defined in the previous chapters.

Circuit Configurations

There are three types of carrier synchronizers known at this time: the *squaring loop*,[3] the *Costas loop*,[4] and the *remodulator*[5] (or *inverse* modulator or *reverse* modulator or *unmodulator*). Block diagrams and operating equations are shown in Figures 11.3 to 11.6.

The squaring loop is the easiest to understand. Its nonlinear element is conveniently modeled as a square-law device, so the output of the nonlinearity is

$$v_x(t) = m^2(t)\sin^2(\omega_i t + \theta_i) = \tfrac{1}{2}m^2(t)\left[1 - \cos(2\omega_i t + 2\theta_i)\right] \quad (11.2)$$

A conventional PLL, operating at double the carrier frequency, locks to the second-harmonic component of (11.2) and the VCO output is divided by 2 to provide the desired reference carrier at the signal frequency. A second-harmonic component exists for any message waveform for which $\mathrm{avg}(m^2) \neq 0$.

Alternatively, let us consider the nonlinearity to be a frequency doubler. Input consists of either of two phasers at $\pm 90°$, as sketched in Figure 11.3. Frequency doubling also doubles the phase of each of the two phasers so that they fall on top to each other at $\pm 180°$, at the doubler output. The

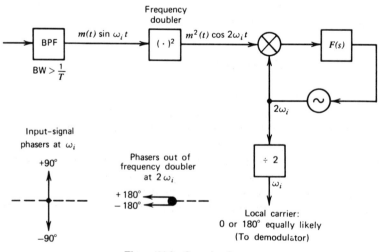

Figure 11.3 Squaring loop.

input phasers, on the average, cancel one another, so there is no carrier component at the input. The output phasers add up in phase to produce a strong second harmonic.

The divide-by-2 following the VCO can operate in either of two phases, determined by the random starting state of the divider. Because of this phase indeterminancy, it is impossible to decide whether the current bit is supposed to be a 1 or a 0 without further information.

This observation introduces us to the fundamental *ambiguity* of all phase-shift modulation techniques; if the information is transmitted in N different phases, there is an N-fold ambiguity in the data recovery. The ambiguity is not a defect of the squaring loop—or other carrier synchronizer—but is inherent to suppressed-carrier, phase-shift keying. It can only be resolved by special encoding or other information carried in the message. Ambiguity-resolution methods are not described here.

Two versions of a remodulator, or inverse modulator, are shown in Figures 11.4 and 11.5; they are minor rearrangements of one another and have identical performance. In Figure 11.4, the incoming signal is demodulated and the message waveform $m(t)$ is recovered. This baseband waveform is used to remodulate the incoming signal; if the waveforms are rectangular and time aligned, then the remodulation removes the modulation completely, so the technique is also sometimes called *modulation removal*. Output of the balanced modulator has a pure carrier component at the input frequency and the PLL tracks that component.

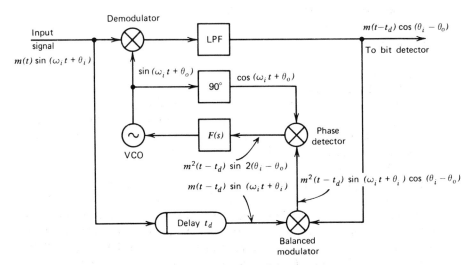

Figure 11.4 Remodulator.

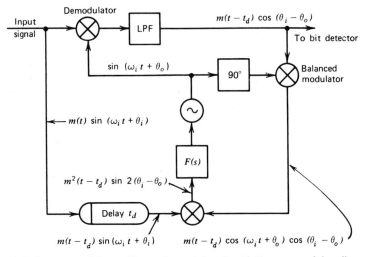

Figure 11.5 Inverse modulator. (Terms "remodulator" and "inverse modulator" are used interchangeably and indiscriminantly.)

The version of Figure 11.5 is slightly different in that the remodulator imposes the recovered message modulation on the VCO output so that both inputs to the phase detector are identically modulated. When taking the low-frequency product of two such waveforms, we find a DC component of the same amplitude as if the inputs had no modulation.

The demodulator, the balanced-modulator or remodulator, and the phase detector are all modeled as ideal multipliers. Some factors of 2 have been neglected in the equations given in Figures 11.3 to 11.5. Often the same circuit element is used in all three places, for example, a diode-ring modulator (Figure 6.9.)

Note that the DC output from the phase detector is proportional to $\sin 2(\theta_i - \theta_o)$; there are two stable tracking points for each cycle of input.

Following the demodulator there is a lowpass filter (LPF) that passes the baseband recovered signal and rejects noise and double-frequency mixer products. Unless there is adequate bandpass filtering prior to the demodulator, it is essential to use a LPF to keep the baseband output well above noise. (The filter, whether bandpass or lowpass, must have sufficient bandwidth to pass the message without excessive distortion.)

Any physical filter necessarily has delay, denoted t_d. The waveforms multiplied together at the balanced modulator (Figure 11.4) or the phase detector (Figure 11.5) must be time aligned. Otherwise the respective modulations are not well-correlated and the available DC output of the phase detector is reduced. If time misalignment is as great as one full pulse interval, the correlation falls to zero and no useful output is available from the PD.

To compensate for the filter delay, a fixed-delay t_d is inserted into the signal path before the signal is applied to the multiplier. Delay placement is shown in Figure 11.4 and 11.5. Besides compensating for the filter delay, it is necessary that $\omega_i t_d = k\pi$ for the circuit phasing be correct, where k is an integer.

A block diagram of a Costas loop is shown in Figure 1.6. It can be derived from a series of passband-to-baseband transformations on a re-modulator circuit. The nonlinear elements are all modeled as ideal multipliers; diode rings, or the equivalent, can be used at the two input multipliers, but a well-balanced baseband circuit is needed for the third multiplier.

Qualitative operation can be visualized as follows: In the absence of modulation, the Q arm acts as a conventional PLL, with the usual PD error voltage being developed at the output of the Q-arm multiplier. When modulation is present, the polarity of the Q-arm output reverses each time that the modulation changes sign; average output is zero (for random data) irrespective of the phase error.

The I-arm multiplier produces a signal in quadrature to that from the Q arm. If phase is nearly correct for tracking, then the I-arm output is the data message. This is used for reversing the Q-arm voltage in the third multiplier, thereby wiping out the polarity reversals that invalidated the Q-arm output as an error voltage. Third-multiplier output is proportional to $\sin 2(\theta_i - \theta_o)$, just as in the remodulator or squaring loop.

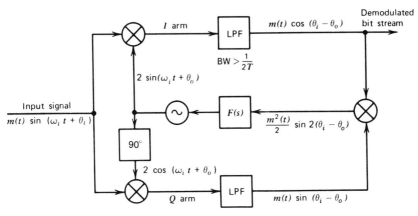

Figure 11.6 Costas loop.

Multiphase Synchronization

All circuits considered so far have been restricted to binary modulation, but modified versions of the three circuits are used for multiphase communications formats. Four-phase examples are shown in Figures 11.7 to 11.9. Technology is still evolving for circuits dealing with more than four phases.

A $\times N$ multiplier is shown in Figure 11.7. Its operation is understood, for rectangular pulses, by simple extension of the squaring loop. For rounded pulses, the desired coherent Nth harmonic is regenerated by intermodulation among the data sidebands. At least Nth-order intermodulation is required to produce the carrier. If the message-waveform bandwidth is sufficiently restricted, then the circuit may fail to regenerate a carrier on some data patterns. (One such pattern would cause the signal phase to advance continually by $1/N$ cycle for each symbol; the circuit interprets this as an unmodulated signal displaced by $1/NT$ Hz from the correct carrier frequency, where T is the symbol interval.)

Block diagrams of four-phase remodulators and Costas loops are shown in Figures 11.8 and 11.9. In a four-phase transmission, independent messages $x(t)$ and $y(t)$ modulate the two quadrature components of the carrier so the transmitted signal can be represented as

$$v_s(t) = x(t)\cos(\omega_i t + \theta_i) - y(t)\sin(\omega_i t + \theta_i) \qquad (11.3)$$

After the usual multiplications and trigonometric identities, the low-frequency output of the phase detector for either the remodulator or the

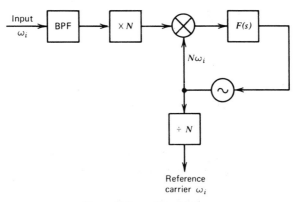

Figure 11.7 $\times N$ multiplier.

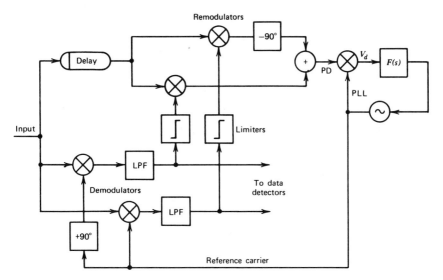

Figure 11.8 Four-phase remodulator.

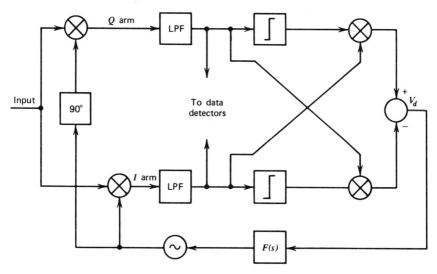

Figure 11.9 Four-phase Costas loop.

Costas loop (they are the same) is found to be

$$v_d(t) = \left[x \sin\theta_e + y \cos\theta_e \right] \mathrm{sgn}\left[x \cos\theta_e - y \sin\theta_e \right]$$

$$- \left[x \cos\theta_e - y \sin\theta_e \right] \mathrm{sgn}\left[x \sin\theta_e + y \cos\theta_e \right] \qquad (11.4)$$

where $\theta_e = \theta_i - \theta_o$ and sgn( ) is the hard-limiting operation.

If the message waveforms are rectangular, the average DC outputs can be calculated as

$$
\begin{aligned}
V_d &= \sin\theta_e & -45° &< \theta_e < +45° \\
&= -\cos\theta_e & 45° &< \theta_e < 135° \\
&= -\sin\theta_e & 135° &< \theta_e < 225° \qquad (11.5) \\
&= \cos\theta_e & 225° &< \theta_e < 315°
\end{aligned}
$$

This characteristic is illustrated by the solid curve of Figure 11.10. It approaches very close to being a sawtooth.

If bandwidth is limited, the pulses cannot be rectangular and the near-sawtooth of Figure 11.10 is modified. Calculations are difficult for

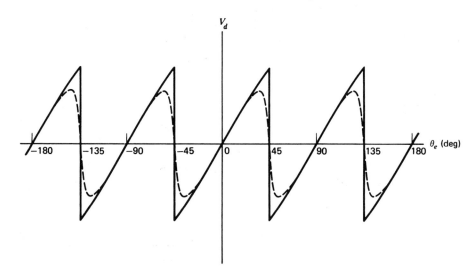

Figure 11.10   Phase-detector characteristic of four-phase remodulator or Costas loop.

rounded pulses (very difficult if the pulses overlap, as they do in a narrowband transmission), but the tendency is for the discontinuities and sharp peaks of the sawtooth to be rounded off, as shown by the dashed curve of Figure 11.10. Noise also causes rounding of the sawtooth: a reversion to sinusoidal as explained in Chapter 6.

A stable lock can be achieved at any of four different phases: 0, ±90, or 180°. There is an inherent fourfold ambiguity that must be resolved by other means.

The limiters shown in Figures 11.8 and 11.9 are essential for the correct operation of the circuits. If linear amplifiers were substituted for the limiters, the circuits would fail to regenerate the desired local carrier.

To regenerate a carrier from a suppressed-carrier, four-phase signal requires at least a fourth-order nonlinearity in the regenerator. Lacking a nonlinear device at the position of the remodulator or Costas limiters, there would be only second-order nonlinearity in the circuit, from the multipliers, and no carrier could be obtained.

A third-order, or any higher odd-order, baseband nonlinearity could be substituted, but the limiter is the easiest to construct.

The Costas loop and remodulator, either two-phase or four-phase, include the data demodulator as part of the circuit. A demodulator is entirely separate if a $\times N$ multiplier is used instead.

In essence, both the remodulator and the Costas loop remove modulation by multiplying by the demodulated message waveform in analog form. Better noise rejection is possible if the message value is optimally detected and the digital message value is used for the modulation-removal multiplication. This type of synchronizer is said to be *decision directed** and has less noise-caused jitter of the reference carrier because the operation of bit detection rejects noise better than the simpler analog-multiplication circuit[6,7].

Additional delays are needed to compensate for the delay of the data-detection circuit; delays must be inserted into the Costas loop as well as the remodulator. Note that the delays and the demodulator lowpass filters are within the feedback loop and must be taken into account when considering stability. Note also that the remodulator really has two feedback loops, which makes for interesting problems in analyzing the dynamic performance.

A decision-directed circuit cannot acquire the carrier until the clock has been acquired and may not be able to acquire the clock until the carrier has been acquired. Decision-directed synchronizers may not be acceptable if fast acquisition is required.

---

*Alternate terminologies are *data aided* and *decision feedback*.

Carrier Synchronizer Performance

In general, analysis of a synchronizer is very difficult because of the presence of the nonlinear regenerator. An example analysis is carried out in Appendix B for the simplest case: a squaring loop with an ideal square-law nonlinearity. The results obtained there are used to point up features that are common to all carrier synchronizers.

Under certain restrictive conditions,* the noise-caused VCO jitter of a squaring loop is approximated by (B.23)

$$\overline{(2\theta_{no})^2} = 4\left(\frac{2B_L N_o}{V_s^2}\right)\left(1 + \frac{N_o B_i}{V_s^2}\right)$$

$$= 4\left(\frac{2B_L N_o}{V_s^2}\right)\left(1 + \frac{1}{2\rho_i}\right) \quad \text{rad}^2 \qquad (11.6)$$

where the recovered carrier phase is θ_o, the VCO phase in a squaring loop is $2\theta_o$, N_o is the spectral density of white, gaussian input noise in V^2/Hz, V_s is the amplitude of the input signal, B_i is the bandwidth of the input bandpass filter in hertz, $2B_L$ is the double sideband noise bandwidth of the PLL, and $\rho_i = V_s^2/2N_o B_i$ is the signal-to-noise ratio in the input filter.

The quantity within the first set of parentheses is recognized as the jitter variance of an ordinary PLL from (3.18). Jitter variance of a squaring loop is $4(1 + 1/2\rho_i)$ times as large as that of an ordinary PLL tracking a pure carrier of the same amplitude in the same noise. The factor of 4 in the variance (it is only $\times 2$ in the rms jitter) arises because of the phase magnification effect of the frequency doubler; the amplitude of any phase disturbance that occurs at the fundamental frequency is doubled at the second harmonic.

Phase magnification appears only in the PLL at $2\omega_i$; it does not affect the recovered carrier phase at ω_i directly. In the squaring loop, the VCO frequency is divided by 2, thereby removing the magnification from the recovered carrier. But the PLL must track the magnified phase; its locking ability is necessarily impaired by the magnification. Where a simple PLL might be able to hold lock down to 0-dB loop SNR, a squaring loop can be expected to lose lock around +6 dB.

[In terms of (11.6), lock will probably fail if $\overline{(2\theta_{no})^2}$ exceeds approximately 0.5 rad^2. Appreciable cycle slipping is incurred for SNRs above loss-of-lock.]

---

*Do not apply (11.6) before reading Appendix B, which explains the conditions.

In the second set of parentheses we see an increase of jitter caused by *squaring loss*. This term arises because of intermodulation of noise by itself in the squarer; at low input SNR, the noise-times-noise term comes to dominate. No such term exists in the simple PLL of Chapter 3. In the squaring loop, the jitter increases by 2 dB for each 1 dB of SNR reduction at small enough levels of input SNR. This phenomenon occurs even for small jitter, with the PLL within its linear region; it is inherent to the nonlinear regenerator and is not caused by the PLL per se.

Identically the same effects occur for a Costas loop or a remodulator. Although VCO jitter in these two configurations is the same as recovered-carrier jitter, the PD characteristic is proportional to $\sin 2\theta_e$, so the PD has only half the dynamic range of the conventional PLL. Loss of lock can be expected in the vicinity of $SNR_L = +6$ dB.

In an N-phase synchronizer, the jitter can be represented as

$$\overline{(N\theta_{no})^2} = N^2\left(\frac{2B_L N_o}{V_s^2}\right)L_N(\rho_i) \tag{11.7}$$

where N^2 shows the N-fold phase magnification and $L_N(\rho_i)$ is the loss caused by noise intermodulation. Phase magnification elevates loss of lock by $20\log_{10}N$ dB as compared to a simple PLL.

Intermodulation loss depends on N, ρ_i, and the nature of the regenerator nonlinearity. There can be a substantial difference between different nonlinearities.[8,9] Some typical losses are given in Table 11.1 for the special case of an Nth-law nonlinear element.[9]

Examples of jitter from (11.7) are plotted in Figure 11.11 for $2B_L = B_i/1000$, a rather narrow choice of loop bandwidth, but not atypical. The curves are parallel at large input SNR; their vertical separation in that region arises from the $\times N$ phase magnification. Upward curvature at small SNR is caused by the noise intermodulation. Clearly both effects

Table 11.1   Intermodulation Losses of Nth-Law Regenerators

N	$L_N(\rho_i)$
1	1
2	$1 + \dfrac{1}{2\rho_i}$
3	$1 + \dfrac{2}{\rho_i} + \dfrac{2}{3\rho_i^2}$
4	$1 + \dfrac{9}{\rho_i} + \dfrac{6}{\rho_i^2} + \dfrac{3}{2\rho_i^3}$

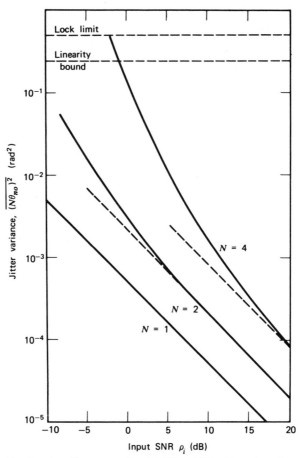

Figure 11.11 Synchronizer jitter example. ($B_i/2B_L = 1000$, N = order of nonlinearity in regenerator.)

combine to degrade tracking severely as compared to the conventional PLL. In particular, a synchronizer is not able to hold lock nearly so far down into the noise as can a simple carrier-tracking loop of the same bandwidth.

Synchronizer performance improves if input signal-to-noise ratio ρ_i can be increased. A designer has the synchronizer filters—the bandpass filter for a $\times N$ multiplier or the baseband filters for a Costas loop or remodulator—available as a design parameter. If filter bandwidth is too large, excess noise is admitted and ρ_i is unnecessarily small. If filter bandwidth is too small, the filter cuts off too much of the useful signal spectrum and ρ_i

is again too small. There is an optimum choice of filter that will maximize ρ_i.

Layland[10] has shown that the optimum filter approaches the signal-matched filter asymptotically for small ratios of signal density to noise density. That is, if we let the pulse wave shape received at the filter input be denoted by $p(t)$ and let its Fourier transform be $P(j\omega)$ and the input noise is white, then the optimum filter $Y(j\omega)$ is approximated* by $|Y(j\omega)|^2 = |P(j\omega)|^2$. This result applies directly to the baseband filters of the Costas loop or remodulator and the bandpass equivalent applies to a $\times N$ synchronizer. Just exactly this filter is often optimum for data detection, so the same filter serves both purposes—synchronization and detection—simultaneously and optimally.

In many digital communications systems, multiple, identical channels are close in frequency. The equipment must be designed to avoid *adjacent-channel interference* (ACI). Experience and simulations have demonstrated that ACI must be strongly excluded from the nonlinear regenerator of the carrier synchronizer or else significant performance degradation will be incurred.

Regenerators in the Costas loop and remodulator are protected against ACI by baseband filters, while a $\times N$ multiplier relies on the bandpass filter. The fact that baseband filters are often easier to build might be one reason to prefer a baseband configuration to the $\times N$ multiplier, particularly if the carrier frequency is very large compared to the symbol rate.

Numerous papers[6,7,11–17] provide alternate and deeper analyses of carrier synchronizers. By applying Fokker-Planck methods, it is possible to obtain the probability density of the jitter and cycle-slip statistics. Probability density, in turn, can be used to calculate the effect on data error probability. Most of the literature is specialized to rectangular signaling pulses.

As a concluding remark, be warned that a carrier synchronizer may be able to lock to data-related sidebands at frequencies spaced k/NT Hz from the carrier[18–21] ($k = \pm 1, \pm 2,...$). If the initial carrier frequency uncertainty prior to lock includes any of these side frequencies, the reference output might very well get locked to the sideband instead of the desired carrier frequency.

---

*More precisely, if $\Phi_s(\omega)$ is the spectral density of the input signal prior to filtering, the SNR at the output of the squarer is maximized if the bandpass filter is

$$|Y(j\omega)|^2 = \frac{\Phi_s(\omega)}{\Phi_s(\omega) + N_o/2}$$

Computer simulations indicate that jitter is not very sensitive to moderate departures from the optimum filter.

The origin of sideband lock is most easily seen in a squaring loop, although the cause is essentially the same in any configuration. Output of the squarer is $m^2(t)\cos 2\omega_i t$, after the DC terms are discarded. Either because of the filtering or because of the way it was originally generated and transmitted, $m(t)$ does not have a rectangular wave shape. In consequence, the envelope of the signal applied to the squarer is not constant but varies with the data pattern (See Figure B.1.) It can be demonstrated[21] that $m^2(t)$ contains discrete spectral lines at DC and at harmonics of the symbol rate.

The DC component multiplies $\cos 2\omega_i t$ to produce the desired double-frequency carrier, while the symbol-rate components produce coherent sidebands spaced $1/T$ apart. Given sufficient signal-to-noise ratio, the PLL can lock to any of these sidebands. After division by 2, frequencies of the reference carrier lock are spaced by $1/2T$. (For an N-phase system, the reference carrier locks are spaced by $1/NT$.)

Some protection against sideband locks is obtained by placing a hard limiter between the bandpass filter and the squarer. A perfect limiter would eliminate the data-related envelope variations.

Two difficulties prevent complete effectiveness of a limiter. Firstly, the envelope variations include complete nulls for each phase reversal (Appendix B). Only a perfect hard limiter can limit on vanishingly small input, and any physical circuit is less than perfect. Therefore, the envelope variation cannot be suppressed completely.

Secondly, if the signal is not centered in the filter passband there is AM—PM conversion due to asymmetrical transmission through the filter (see Appendix B). The PM sidebands so generated are not affected by a limiter.

Another antisideband approach is to utilize a frequency discriminator to aid correct acquisition (Chapter 5). The resemblance of a Costas loop to a quadricorrelator is marked (compare Figures 5.14 and 11.6). The two detector circuits can be combined so as to share components. In fact, simply using different filters in the two arms of the Costas loop produces a frequency-aiding constituent in the loop error voltage.[21]

11.3 CLOCK SYNCHRONIZERS

In contrast to the small number of carrier synchronizer types, there is a very large number of clock synchronizer configurations, and new ones appear regularly. Several examples are chosen to illustrate the operation of clock synchronizers. Although enough is known of the subject to permit design of very good circuits, a better understanding is still needed.

Bandwidth Regimes

Clock synchronizers can be categorized according to the bandwidth of the communications system. We distinguish *wideband* and *narrowband* signaling regimes. Clock-synchronizer designs for the two regimes are significantly different: they face different problems; there have been two disjoint sets of investigators; and the authors of the existing literature are divded into two camps. Examples of papers dealing with wideband clock recovery may be found in Refs. 23 to 29 while narrowband examples are in Refs. 22 and 30 to 36.

We assume that the digital information is transmitted by weighted pulses, each with identical shape and spaced uniformly by T secs. Information is carried in the weighting (e.g., ± 1 weights for binary transmission). This section deals with baseband pulses, but it is to be understood that they might very well have been retrieved by demodulating a passband signal.

The characteristics of the two regimes are given in Table 11.2.

Signal Representation

A baseband data pulse stream may be represented as

$$\sum_n a_n p(t - nT) \qquad (11.8)$$

where a_n is the data weighting applied to the nth pulse and $p(t)$ is the shape of each pulse. If the data are binary, $a_n = \pm 1$. The a_n take on more than two values if the data are nonbinary.

Maximum Likelihood Trackers

If the pulse waveform is strictly confined to the interval T [i.e., $p(t)=0$ for t outside the interval $(0, T)$], then there is an optimum clock timing known variously as the maximum likelihood (ML) estimate[23-25] or the maximum *a posteriori* (MAP) estimate.[26] The usual derivation of the optimum estimate assumes an open-loop search and measure implementation, which impractically assumes that the phase holds stable once the optimum estimate has been found. Also, the derivation does not provide a ready evaluation of the errors in the estimate.

Ordinarily it is necessary to use a closed-loop tracking synchronizer to accomodate relative phase drift between the incoming signal and the local clock. Stiffler[24] and Mengali[25] have devised tracking synchronizers that

Table 11.2   Signaling-Bandwidth Regimes

Wideband Regime

Bandwidth occupancy greatly exceeds the signaling rate $1/T$

Signaling pulses are essentially confined to single symbol interval T. Pulses may be, but need not be, rectangular

Dominant disturbance is usually white, additive, gaussian noise. In some applications, jamming or other cochannel interference may be dominant

Coding may be used on the digital message, implying that input SNRs might be quite low

Examples of wideband links may be found in military and deep space communications

Narrowband Regime

Bandwidth occupancy is very small, approaching the Nyquist limit of $1/2T$ Hz in baseband or $1/T$ Hz in double-sideband modulated signals

Pulse shapes spread over many symbol intervals T; pulses overlap. Rectangular shape is not even a good approximation

Dominant disturbance is often from overlapping pulses, known variously as *intersymbol interference* (ISI) (in communications systems) and *pulse crowding* (in playback of digital magnetic recordings). Signal-to-noise ratios are often large (e.g., 30 dB minimum is typical in some telephone-line applications), although small SNRs are also encountered

converge to the ML estimate and which, it is hoped, should have much the same phase error statistics. (Because the mechanizations of the open- and closed-loop synchronizers are so very different, it is impossible that they would have identical statistics. If the errors are small—the condition of most practical interest—we would hope that the error variances would be nearly the same.)

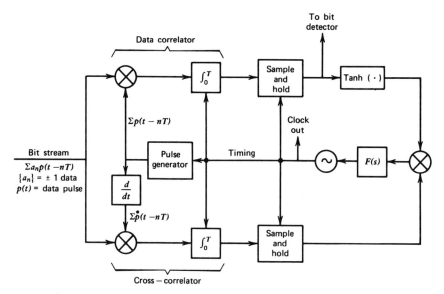

Figure 11.12 Clock synchronizer based on maximum-likelihood estimate.

Mengali's tracker is shown in Figure 11.12; there is a distinct resemblance to a Costas loop.* In the upper arm the incoming signal is correlated, symbol-by-symbol, with a stored replica of the signal pulse shape. The correlator is sampled at T-sec intervals and the integrator in the correlator is discharged ("dumped") immediately after each sample.

(Once the timing is properly aligned, the upper correlator provides optimum filtering for detection of the data value. The correlator samples can be used for data decisions.)

In the lower arm, the incoming signal is correlated against a stored replica of the derivative of the signal pulse shape. Output of this correlator is also sampled at T-sec intervals and the integrator is then dumped instantaneously in preparation for the next integration interval. When the local timing agrees with that of the incoming signal, the average output of the lower correlator is zero; when there is a timing error, the lower correlator output departs from zero.

If the incoming pulses were all of the same polarity, the polarity of the lower correlator would show the direction of the timing error. However, the incoming data have random signs, so the lower correlator has random polarity of samples.

---

*This resemblance is not accidental; the Costas loop is an approximation to a tracking ML estimator of carrier phase.[8]

The samples from the upper arm contain the data-sign information, so multiplying the two arm outputs together removes the data randomness and provides a signal containing error-sense information. This is then treated in the same manner as the error voltage in an ordinary PLL or a Costas loop; the loop is closed through the usual loop filter and VCO and, for this synchronizer, a pulse generator as well.

The tanh nonlinearity arises out of the ML derivation and is approximated by a soft limiter. In fact, if SNR is large, it is well approximated by a hard limiter, while for small SNR the approximation is best if the path is linear.

Mengali has analyzed the circuit of Figure 11.12 for white noise, for $p(0) = p(T) = 0$ (an important restriction, as we see later), for large SNR, for $B_L T \ll 1$, and for a hard limiter. His result is

$$\sigma_\varepsilon^2 = \frac{T B_L}{rb} \tag{11.9}$$

where $\varepsilon = \tau/T$, τ is the timing error, $r = E_p/N_o$ is the input SNR defined in data terms, N_o is the one-sided noise density, E_p is the energy per signal pulse, $b = T^2 E_{\dot{p}}/E_p$, and $E_{\dot{p}}$ is the energy of $\dot{p}(t) = dp(t)/dt$.

$$E_p = \int_0^T p^2(t)\, dt$$

$$E_{\dot{p}} = \int_0^T \dot{p}^2(t)\, dt$$

By use of a raised-cosine signal pulse, the results of (11.9) were compared against a report of Monte Carlo simulations[23] for the maximum-likelihood estimator. The tracker phase variance was roughly 20% worse than optimum, a minor degradation. Analytical prediction of the optimum is difficult and there does not appear to be any numerical evaluation other than Ref. 23.

For engineering purposes, (11.9) might be regarded as a lower bound on the achievable jitter from a tracking clock synchronizer. Furthermore, Figure 11.12 provides a guide to the intelligent implementation of a wideband synchronizer.

Mengali's analysis breaks down for a rectangular pulse shape since $p(0) \neq 0$ and $p(T) \neq 0$. The ML timing estimate for a rectangular pulse exists, but the circuit of Figure 11.12 cannot find it. This is unfortunate because of the practical importance of rectangular signal pulses. Other *ad hoc* schemes must be employed.

Early-Late Gates

An early gate–late gate synchronizer is very popular for rectangular pulses; one example is shown in Figure 11.13. Many variations are possible[24,26,27]; the example shown here is one of the easiest to explain, has been employed frequently, and is convenient to mechanize.

The heart of the circuit consists of a pair of gated integrators, each performing its integration over a time interval of $T/2$ sec. Integration by the early gate occurs in the $T/2$ sec preceding the nominal location of data transitions, while the late gate integrates during the $T/2$ sec immediately following the transitions. Gate intervals adjoin one another but do not overlap. If timing error is zero, the data transitions fall exactly on the boundary between early and late gates. Illustrative waveforms are shown in Figure 11.14.

If timing error is not zero, a transition falls not on the boundary but within one or the other of the gates. Since signal polarity changes within the gate containing a transition, the associated integration reaches a lesser magnitude than when the transition is external to the gate. Comparison of magnitudes of the two integrations gives an indication of timing error.

We denote the timing error by τ ($|\tau| \leqslant T/2$). Useful gate output is the integrated value that has been accumulated after $T/2$ sec; that is, the

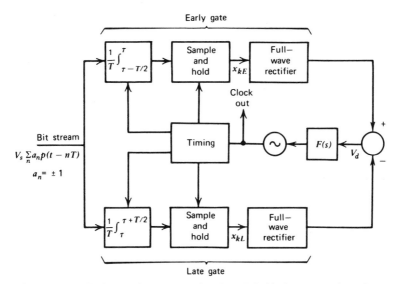

Figure 11.13 Early gate–late gate synchronizer. Suitable for rectangular pulses.

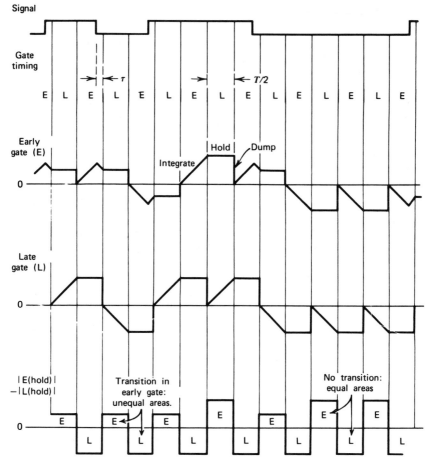

Figure 11.14 Waveforms of early–late gate (rectangular input).

integrator is sampled after the integration has been completed. The loop does not see the integration details that precede the sample. We designate the samples as x_{kE} and x_{kL}, where k is a counting index on a sequence of data pulses. Sample values are

$$x_{kE} = \frac{1}{T_I} \int_{\tau + \left(k - \frac{1}{2}\right)T}^{\tau + kT} s(t)\,dt; \quad x_{kL} = \frac{1}{T_I} \int_{\tau + kT}^{\tau + \left(k + \frac{1}{2}\right)T} s(t)\,dt \quad (11.10)$$

where T_I is the time constant of the integrators and the incoming binary signal is

$$s(t) = V_s \sum_n a_n p(t - nT) \tag{11.11}$$

The pulse $p(t)$ is taken as rectangular, with duration T.
For $\tau \geqslant 0$, the integration sample values are

$$x_{kE} = \frac{V_s}{T_I} \left[\left(\tfrac{1}{2} T - \tau \right) a_{k-1} + a_k \right]$$

$$x_{kL} = \frac{V_s}{T_I} \frac{a_k T}{2} \tag{11.12}$$

Similar forms obtain for $\tau < 0$. In practical analog circuits, the samples are held for a time qT ($q \leqslant 1$)*, their magnitudes are compared and then dumped, and the cycle repeats. From the waveforms of Figure 11.14, it may be seen that the held samples are compared sequentially, not simultaneously; this expedient leads to simple circuitry but causes a strong ripple at the bit rate frequency.

The error from a pair of samples is

$$y_k = |x_{kE}| - |x_{kL}| \qquad \text{(for all } \tau)$$

$$= \frac{V_s}{T_I} \left[\left| \left(\tfrac{1}{2} T - \tau \right) a_{k-1} + a_k \tau \right| - \tfrac{1}{2} T |a_k| \right] \qquad \text{(for } \tau > 0) \tag{11.13}$$

Average error voltage applied to the loop is

$$V_d = q E(y_k) \tag{11.14}$$

where $E(\ )$ indicates statistical average. Sample values depend on data sequence and timing error as shown in Table 11.3

Table 11.3   Error Sample Values ($\tau > 0$)

Data Sequence	$y_k T_I / V_s$
$a_{k-1} = a_k$	0
$a_{k-1} \neq a_k$	-2τ; $(0 \leqslant \tau \leqslant \tfrac{1}{4} T)$
	$2\tau - T$; $(\tfrac{1}{4} T \leqslant \tau \leqslant \tfrac{1}{2} T)$
and similarly for $\tau < 0$	

*In Figure 11.14, $q = 0.5$.

We see that a nonzero error signal is generated only when a transition occurs between bits. This behavior is typical of most practical signal formats and synchronizer types because timing information is carried only in the data transitions. In the absence of a transition, no timing information is present and the voltage ought to be zero.

If there should be a long run of data without transitions, no error information is supplied to the loop for the duration of the run. Phase of the VCO will drift off from its correct position because of noise and circuit offsets. If the run is long enough, compared to the drift rate, the loop eventually slips one or more clock cycles. A clock slip is ordinarily a serious matter since it also destroys frame synchronization; sensible data cannot be retrieved until frame sync has been reacquired once again.

Probability of a transition between bits is labeled d and is often called the *transition density* of the bit stream. For random binary data, $d=0.5$ but long runs without transitions are possible. System designers often must select signal designs that guarantee sufficient transition density, even over fairly short runs.

Extending Table 11.3 to negative timing errors and taking transition density into account yields an average error voltage

$$V_d = \frac{-qV_s d}{T_I}\tau \qquad \left(|\tau| \leqslant \tfrac{1}{4}T\right)$$

$$= \frac{qV_s d}{T_i}\left(\tau - \tfrac{1}{2}T\right) \qquad \left(\tfrac{1}{4}T \leqslant \tau \leqslant \tfrac{1}{2}T\right) \qquad (11.15)$$

$$= \frac{qV_s d}{T_I}\left(\tau + \tfrac{1}{2}T\right) \qquad \left(-\tfrac{1}{4}T \geqslant \tau \geqslant -\tfrac{1}{2}T\right)$$

which is a periodic, triangular characteristic, as sketched in Figure 11.15.

There can be variations on the basic scheme, mainly in the timing of the gates. Shorter gate times can be used—an expedient that might improve noise performance but that introduces dead regions when the timing error is such that the transitions fall into neither gate. An extreme example is shown in Figure 5.4.

If the gates are shorter than $T/2$ they need not adjoin one another. In this case a portion of the dead zone is introduced into the region of the tracking crossover. Dead zones at the tracking null usually impair feedback stability and are to be avoided if at all possible.

Gate intervals longer than $T/2$ seconds are possible; the gates would have to overlap to accomodate the longer durations.[26]

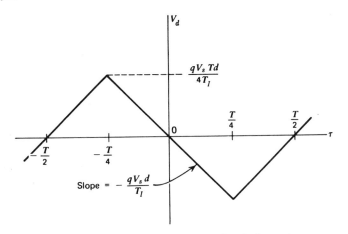

Figure 11.15   Average error output of early–late gate.

Noise analysis of a nonlinear synchronizer circuit is a formidable under-taking. Stiffler[24] has produced a semilinear analysis that includes the effects of squaring loss. Simon has analyzed overlapping gates[27] by Fokker-Planck methods and therefore obtains slip probability as well as error variance. There does not appear to be any complete nonlinear analysis published for the arrangement of Figure 11.13.

By imposing extreme simplifications, it is possible to perform an approximate linear analysis of the same form as that in Chapter 3. The approximations are valid only if the input signal-to-noise ratio is large. This condition is often, but not always, attained in practical communications systems.

Let us consider the individual gated integrators, where superposition of signal and noise still applies; that is, the contributions of signal and noise to the integrated values can be considered separately and merely summed together. Input to the integrators is $s(t) + n(t)$, where $s(t)$ is defined in (11.11) and $n(t)$ is white, gaussian, zero-mean noise of one-sided spectral density N_o V^2/Hz.

Following Papoulis[37] the noise variance of integrated samples is

$$E\left(x_{nkE}^2\right) = E\left(x_{nkL}^2\right) = \frac{N_o T}{4 T_I^2} \tag{11.16}$$

where x_{nk} represents the noise contribution to the kth sample, either early or late, as indicated by the final subscript. We assume that timing error is zero, so the magnitude of the signal integration is $V_s T / 2 T_I$ for all samples.

Each sample is applied to a full-wave rectifier whose output will be $|V_s T/2T_I + x_{nk}|$.

For a large enough signal-to-noise ratio, the first term almost always exceeds the second, so the rectifier output can be approximated as $V_s T/2T_I + x_{nk} \operatorname{sgn} x_k$, where x_k is the data value within the gate interval. By making this approximation, we neglect the squaring loss inherent to all rectifiers. At low signal-to-noise ratio the loss is not negligible so the analysis would no longer be valid.

We define $\tilde{x}_{nk} = x_{nk} \operatorname{sgn} x_k$, which is really the noise samples with random (data-related) reversals of polarity. Since the signal and noise are independent of one another, statistics of $\tilde{x}_{nk}$ are identical to those of x_{nk}. In particular, $E(\tilde{x}_{nk}^2) = E(x_{nk}^2) = N_o T/4T_I^2$.

Signal components of the integrator output tend to cancel (they should cancel exactly if timing error is zero), while noise components are completely independent and add on a power basis. Phase detector output can be represented as

$$v_{dk} \approx -\frac{qV_s d}{T_I}\tau + qn_k + \text{ripple} \qquad (11.17)$$

where $n_k = \tilde{x}_{nkE} - \tilde{x}_{nkL}$; henceforth ripple is neglected. Ripple must be dealt with by the designer and can be troublesome in actual practice. It does not enter into the analysis of additive noise effects.

We define $n_k' = \dfrac{n_k T_I}{V_s d}$ so

$$v_{dk} = \frac{qV_s d}{T_i}(-\tau + n_k') \qquad (11.17a)$$

Note that n_k' has dimensions of time and may be regarded as an additive source of timing error introduced into the loop in a manner analogous to the introduction of n' of Chapter 3.

From its definition, we know that n_k' has zero mean and its variance is

$$E(n_k'^2) = \left(\frac{T_I}{V_s d}\right)^2 E\left[(\tilde{x}_{nkE} - \tilde{x}_{nkL})^2\right] = \left(\frac{T_I}{V_s d}\right)^2 \left[E(\tilde{x}_{nkE}^2) + E(\tilde{x}_{nkL}^2)\right]$$

$$= \frac{N_o T}{2V_s^2 d^2} \quad \sec^2 \qquad (11.18)$$

The following noise properties were used in obtaining the result:

$$E(\tilde{x}_{nkE}\tilde{x}_{nkL}) = 0 \qquad E(\tilde{x}_{nk}\tilde{x}_{nj}) = 0 \quad (\text{for } k \neq j)$$

$$\tilde{x}_{nk}^2 = (x_{nk} \operatorname{sgn} x_k)^2 = x_{nk}^2$$

Noise samples occur at regular intervals of $T/2$ sec. Samples are independent, so their spectrum is flat over the Nyquist band 0 to $1/T$ Hz. Spectral density of the noise samples is

$$\Phi_{n_k'} = \frac{E(n_k'^2)}{(1/T)} = \frac{N_o T^2}{2V_s^2 d^2} \qquad (\text{sec})^2/\text{Hz} \qquad (11.19)$$

This noise propagates through the loop, is filtered by the closed loop transfer function, and causes a time jitter $\Delta\tau$ of the VCO output.

As in Chapter 3, the time jitter can be approximated as

$$E[\Delta\tau^2] = \Phi_{n_k'} B_L \qquad \text{sec}^2 \qquad (11.20)$$

The approximation assumes $B_L T \ll 1$.

It is convenient to normalize the timing jitter to the bit interval whence

$$\boxed{E\left[\left(\frac{\Delta\tau}{T}\right)^2\right] = \frac{N_o B_L}{2V_s^2 d^2}} \qquad (11.21)$$

is the variance of fractional timing error introduced by the additive noise.

Transition-Tracking Loop

The early–late gate loop acts to place the boundary between the two gates exactly at the transition instant. Departure from that timing generates the loop error signal.

It is also possible to produce an error signal by using a single gate that straddles a transition. If the transition is exactly centered within this *midphase* gate, integration over the gate interval is zero. If the transition is not centered, then the integration produces a positive or negative error output.

Sense of the error must be determined according to the direction of the transition. Also, if there is no transition, then no information has been presented and the integrator output must be ignored. Transition detection and direction sensing can be performed by conventional detection of the data bits (in an *inphase* integrate-and-dump circuit) and by digital logic operations on the current detected bit and its predecessor.

Lindsey and Tausworthe[26,28] invented a scheme that implements the operations outlined above and have called it the *digital data-transition tracking loop*: " digital", because it was built largely from digital hardware. A complete description can be found in Refs. 26 and 28 and a detailed nonlinear analysis is in Ref. 29. Characterization of operation is far better than for the early–late gate. It is not clear that either method has significant performance advantage over the other when both are optimized.

Other Wideband Synchronizers

The delay-line multiplier of Figure 11.16 is attractive in its simplicity, provided that the input signal-to-noise ratio is large. Analysis shows that the output of the multiplier contains a discrete spectral component at the clock frequency. Amplitude of the regenerated clock depends on the pulse waveform and the delay time t_d. For rectangular pulses, it can be shown that the clock-component amplitude is maximized if $t_d = T/2$.*

Not only can the delay-line multiplier regenerate a clock from a baseband bit stream, but it also works on a passband PSK signal. Output amplitude is proportional to $\cos \omega_i t_d$, where ω_i is the carrier frequency of the incoming signal. Frequency and delay must be chosen such that $|\cos \omega_i t_d| \approx 1$. Otherwise the amplitude of the regenerated clock is reduced; particularly destructive effects occur if $|\omega_i t_d| = \pi/2$.

Examination of waveforms of the delay-line multiplier would show that its output consists of a unipolar pulse (rectangular, of duration $T/2$, if the signaling pulses are rectangular and $t_d = T/2$) at the location of each data transition and is zero otherwise. Noise is neglected in that description. Spectral analysis[22] of the resulting pulse train yields the clock-rate component.

The delay-line multiplier is just one form of *transition detector*. Any circuit that generates a unipolar pulse at each signal crossing is a transition detector. The resulting sequence of unipolar pulses will contain a clock-rate component. Figure 11.17 shows a couple of transition detectors that have seen use in various applications.

In recovery of bit timing from magnetic recording media—tapes, discs, and drums—the signal-to-noise ratio is usually large, so additive noise has little influence. One common scheme is to hard limit the playback signal and utilize a special form of sequential phase detector (Chapter 6) to determine the timing error. Operation of the sequential PD must be such that output is zero in the absence of a transition. In particular, a phase-frequency detector is rarely applicable.

Sequential phase detectors are notorious for fractional-harmonic locks (Chapter 10); the designer should be alert to the possibility of their occurrence.

Playback speed of tape machines is not constant; the data stream is frequency modulated by speed variations known as *flutter*. To detect the

*If pulses are sufficiently bandlimited—specifically, Nyquist pulses with no energy outside the band 0 to $1/T$ Hz—then the clock amplitude is maximized for $t_d = 0$.

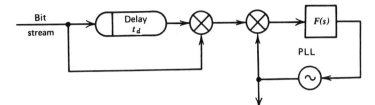

Figure 11.16 Delay-line multiplier. This system also works with PSK IF input.

data properly, it is necessary that the local recovered clock have much the same flutter as the data stream. Therefore, the clock synchronizer needs to have sufficient bandwidth to track most of the flutter modulation. If the synchronizer fails to track the flutter—if the loop bandwidth is too small —then the flutter appears as timing error in the data detection. For high data rates, the data phase excursions could easily correspond to several symbol intervals on typical machines, which would be disastrous if not tracked out.

All the synchronizers shown in this chapter—clock and carrier, wideband and narrowband—are implemented in analog circuits. Digital implementation is also possible and may even be preferable, especially for

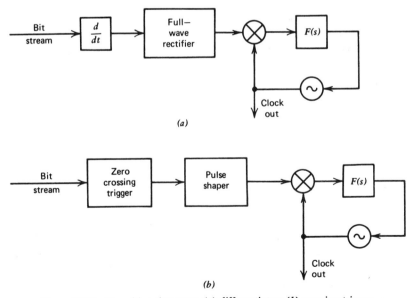

Figure 11.17 Transition detectors: (a) differentiator; (b) crossing trigger.

low-rate clock synchronizers. Examples may be found in Refs. 38 and 39; the latter contains a sizable bibliography.

Narrowband Synchronizers

The outstanding feature of bandlimited pulse trains is that the individual pulses overlap one another. Overlap can extend over many pulse intervals in truly narrowband systems.

Because of the overlap, correlators or gated integrators are unsatisfactory for data decision filtering or for clock regenerator circuits. If gate time is restricted to one symbol interval, the tails of the current pulse are lost and there is interference from tails of neighboring pulses. If gate time spans more than one pulse interval, there is even worse interference from other pulses. Narrowband receivers almost invariably use filters rather than correlators; data detection is accomplished by sampling the filter output.

To avoid *intersymbol interference* (ISI) the pulses are given a Nyquist shaping[40] whereby the tails of one pulse go through zero at the sampling times of all other pulses. Nyquist shaping works very well to eliminate ISI at sample times in the data detection circuits, but it is not sufficient to eliminate the effects of ISI on the clock synchronizer. In general, the recovered clock wave has a jitter component caused by ISI. Since the jitter is caused by the data pattern it has been called *pattern noise* or *pattern jitter*. In many applications it predominates over the additive noise and can be the major concern of a synchronizer designer.

To introduce the problems of narrowband synchronizers, we consider a separate regenerator of the general type suggested in Figure 11.2a. (The same behavior obtains with the nonlinear phase-detector type of regenerator, but the explanation is easier with a separate nonlinearity.) Henceforth we neglect additive noise.

Output of the regenerator can be written as

$$v_r(t) = A \sin \omega_s t + a(t) \sin \omega_s t + b(t) \cos \omega_s t \qquad (11.22)$$

where $\omega_s = 2\pi/T$ is the clock frequency, A is the amplitude of the regenerated clock (to which the PLL locks), and $a(t)$ and $b(t)$ are random, zero-mean disturbances caused by the data and ISI.

The PLL locks to the first term. If tracking is accomplished without phase error (see Chapter 4), then the loop does not see $a(t)$. If there is a phase error θ_e, the PD output contains a term $a(t)\sin\theta_e$ as well as $b(t)\cos\theta_e$. To avoid interference from $a(t)$, the inphase component, the loop phase error should be kept very small.

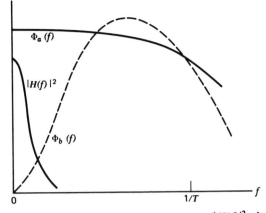

Figure 11.18 Quadrature spectra from clock regenerator. ($|H(f)|^2$ shows transmission through PLL.)

Statistics of $a(t)$ and $b(t)$ are very difficult to determine. No general solution, applicable to all pulse shapes, exists. However, some special cases, simulations, and hardware experience all point to a curious result, which is now discussed.

The nonlinearity spreads the spectra; if the data spectrum extends roughly from 0 to $1/2T$, the spectra of $a(t)$ and $b(t)$ extend from 0 to about $1/T$. This kind of behavior is typical of even-order nonlinearities and is not restricted to clock synchronizers. The in-phase spectrum $\Phi_a(f)$ tends to be an ordinary, lowpass spectrum: a spread and somewhat distorted version of the data spectrum, as in Figure 11.18. The quadrature spectrum $\Phi_b(f)$ is quite different; it is small or zero at zero frequency and rises monotonically as frequency increases, at least out to the high-frequency rolloff.

Transfer function of the PLL plots as a narrowband, lowpass filter concentrated near zero frequency in Figure 11.18. Quadrature density is small and not flat within the loop passband; simple formulas used previously for white noise are not applicable.

One consequence of the shape of Φ_b is that the phase variance it causes rises as a steep function of loop bandwidth. For white noise, the variance would be directly proportional to B_L, but the quadrature pattern jitter can go as B_L^2 or B_L^3, depending on pulse shape and nature of the nonlinearity. Prudence requires that the bandwidth be kept small, even if additive noise is negligible.

Another consequence is that the resulting jitter has high-frequency content, much greater than ordinarily anticipated. One particular, formal analysis[30] predicts that the jitter could be infinite for the usual PLL in which the loop rolloff is only -6 dB/octave. The ill effects of the high-frequency content can be suppressed by including an extra pole in the PLL, presumably at a frequency well beyond the open-loop gain crossover. "Infinite" variance is a manifestation of potential cycle slips, but these are ameliorated by the extra pole.

Faced with the spectra of Figure 11.18, it is evident why the steady loop phase error must be kept small. If tracking error is zero, only the quadrature disturbance contributes to jitter and its density is small within the PLL bandwidth. If tracking error does not vanish, some of the in-phase disturbance is cross-coupled into a quadrature disturbance that generates phase jitter. Since the in-phase disturbance is relatively large within the PLL passband, small amounts of cross coupling can cause large increases of jitter.

Existence of quadrature disturbances means that phase modulation is included in the regenerated clock wave. An alternative view is that the zero crossings of the regenerated clock wave are not equispaced. The latter condition is an inevitable feature of data pulse trains whose bandwidth is too narrow to satisfy Nyquist's second criterion.[40] As a corollary, if the bandwidth is wide enough to avoid pulse overlap, then it should be possible to avoid quadrature disturbances, which is the reason that pattern jitter can be of less concern in wideband systems than in narrowband systems.

We should not conclude that overlapping pulses must necessarily be afflicted with pattern jitter. Firstly, Nyquist's second criterion does not forbid pulse overlap; it only requires some minimum bandwidth conditions. The corollary does not include all pulse trains that meet the second criterion. More broadly, pattern jitter is associated with clock regeneration from data pulses, which must have at least one nonzero sample to convey information. We see below that it is possible to modify the pulse shape to suppress all pattern jitter, but the modified pulse is not the same one used for data detection.

Before pursuing jitter suppression, it is instructive to illustrate some of the nonlinear regenerators used with narrowband systems.

Full-wave rectifiers are commonplace because of their simplicity (Figure 11.19). Input can be a baseband pulse stream or else the envelope variations of a passband signal can be rectified. (Bandlimited data signals always have envelope variations caused by the data modulation. In many practical modulation formats the envelope variations contain a spectral line at the clock frequency—or a related harmonic. A clock waveform is easily regenerated by simple rectification of the incoming passband signal.)

An infinite variety of rectifier characteristics are possible. Most common are the square-law rectifier and the absolute-value "linear" rectifier. Practi-

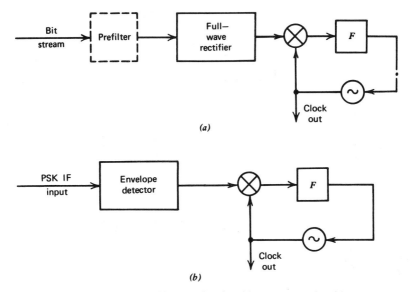

Figure 11.19 Clock rectifiers: (*a*) baseband input; (*b*) passband input.

cal circuits usually fall in between square law and absolute value. Up to 2 dB less jitter is obtained with the absolute-value characteristic, as compared to square law, for most input signal-to-noise ratios of interest.[8,31] The square-law rectifier is prominent in the literature because it is mathematically the most tractable.

The zero-crossing detector of Figure 11.17*b* can also be used as a narrowband regenerator.

Another possibility is the sampled-derivative detector of Figure 11.20, which aligns the strobe pulse with the peak of the signaling pulse. Such alignment is optimum for data detection if the signal pulse is symmetric about its peak; other alignment might be superior for asymmetric pulses.

A modification of the derivative circuit, which separates the regenerator from the phase detector and avoids sampling, is shown in Figure 11.21.

Analysis of zero-crossing and of peak-sampling regenerators may be found in Ref. 30, discussion of square-law rectifiers is given in Ref. 34, and analysis of associated PLL matters is found in Refs. 35 and 36.

Each regenerator has at least a quadratic nonlinearity. Some have higher-order or "harder" nonlinearities. (If the nonlinearity is discontinuous, it cannot be represented by a power series and we cannot speak of the *order* of such a function.) Nyquist baseband pulses occupy a bandwidth $(1 + k)/2T$, where k is known as the *rolloff factor* or *excess-bandwidth*

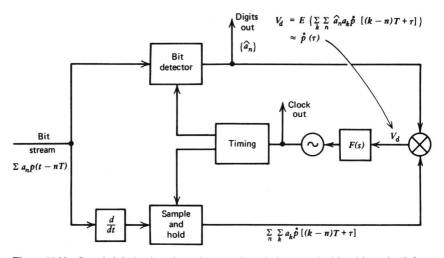

Figure 11.20 Sampled-derivative phase detector. Sets timing to coincide with peak of data pulse.

factor. For a quadratic regenerator, it can be shown that the amplitude of the regenerated clock wave is proportional to k. Therefore, a quadratic regenerator cannot produce a clock from minimum-bandwidth, Nyquist pulses, for which $k = 0$. If k is small—as imposed in some restricted-bandwidth applications—then a quadratic nonlinearity is inadequate.

The square-law rectifier and the simple derivative-product regenerator, without a limiter, of Figure 11.21 have only quadratic nonlinearities. An absolute-value rectifier, a fourth-law rectifier,[32] a derivative circuit with a

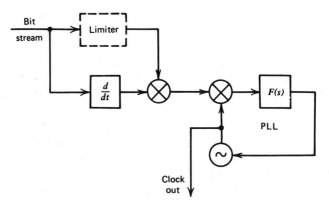

Figure 11.21 Derivative-product regenerator.

limiter or bit detector in one arm, and a zero-crossing detector are all examples of harder nonlinearities.

Each of the example regenerators suffers from pattern jitter for any excess bandwidth less than 100%. Pattern jitter worsens as bandwidth becomes more constricted. The difficulty lies not so much in the circuit itself as in the pulse shapes applied to the nonlinear regenerator. Nyquist pulses are well suited for optimum data detection, but they are poorly adapted for clock recovery. It turns out that modification of the pulse shape by means of a linear filter can suppress the jitter.

Franks and Bubrouski[33] have shown that inserting a suitable prefilter before the clock regenerator can suppress the pattern jitter entirely. Nyquist pulses are delivered to the data detector, but differently shaped pulses are supplied to the regenerator. To suppress pattern jitter, given a square-law rectifier as the regenerator element, requires that the spectrum of the prefiltered baseband pulse train be symmetrical about $1/2T$, half the symbol rate, and that it be bandlimited to the interval $(1/4T, 3/4T)$. It is also necessary that the PLL be bandlimited to $1/T$, which is very large compared to practical bandwidths, and that the PLL operate with zero static phase error to avoid cross-coupling of amplitude fluctuations.

Nyquist shaping provides the high-frequency band limitation needed for the desired spectrum, so a feasible prefilter is a highpass filter to shape the low-frequency portion of the spectrum. Not only does the prefilter suppress pattern noise, but it also reduces the influence of low-frequency components of additive noise. Simulations[31] indicate that quite simple prefilters provide most of the improvement that is theoretically attainable and that complex filters are not necessary.

Prefilters can also be applied to derivative circuits (Figures 11.20 and 11.21) with similar benefits.[31]

A prefilter is successful if all input pulses have the same shape. If different pulse shapes are intermixed or if the channel is nonlinear, thereby causing distortion that varies with data pattern, then a prefilter cannot provide full benefit.

11.4 AFTERWORD

Two final comments are in order:

• This chapter only touches briefly on some high points of synchronizers for digital communications. There remain large areas that have not even been mentioned. A whole book could be written on the subject of synchronizers.

• Much remains to be learned, even concerning the topics considered here.

REFERENCES

1. F. M. Gardner, "Comparison of QPSK Carrier Regenerator Circuits for TDMA Application," *Conference Record*, International Conference on Communications, 1974, Paper 43B.

2. *Transmission Systems for Communications*, revised 4th ed., Bell Telephone Laboratories, December 1971, Chap 27.

3. J. W. Rieke and R. S. Graham, "The L3 Coaxial System: Television Terminals," *BSTJ*, Vol. 32, pp. 915–942, July 1953.

4. J. P. Costas, "Synchronous Communications," *Proc. IRE*, Vol. 44, pp. 1713–1718, December 1956.

5. E. Hopner, "Phase Reversal Data Transmission System for Switched and Private Telephone Line Applications," *IBM J.*, pp. 93–105, April 1961.

6. W. C. Lindsey and M. K. Simon, "Data-Aided Carrier Tracking Loops," *IEEE Trans.*, *COM-19*, pp. 157–169, April 1971.

7. M. K. Simon, "Optimum Receiver Structures for Phase-Multiplexed Modulations," *IEEE Trans., COM-26*, pp. 865–872, June 1978.

8. J. F. Oberst and D. L. Schilling, "The SNR of a Frequency Doubler," *IEEE Trans.*, *COM-19*, pp. 97–99, February 1971.

9. S. A. Butman and J. R. Lesh, "The Effects of Bandpass Limiters on n-Phase Tracking Systems," *IEEE Trans., COM-25*, pp. 569–576, June 1977.

10. J. W. Layland, "An Optimum Squaring Loop Prefilter" *IEEE Trans., COM-18*, pp. 695–697, October 1970.

11. J. J. Stiffler, "The Squaring Loop Technique for Binary PSK Synchronization," *JPL SPS 37-26*, Vol. IV, pp. 240–246, Jet Propulsion Laboratory, Pasadena, CA, April 30, 1964.

12. R. L. Didday and W. C. Lindsey, "Subcarrier Tracking Methods and Communication System Design," *IEEE Trans., COM-16*, pp. 541–550, August 1968.

13. W. C. Lindsey and M. K. Simon, "The Performance of Suppressed Carrier Tracking Loops in the Presence of Frequency Detuning," *Proc. IEEE*, Vol. 58, pp. 1315–1321, September 1970.

14. W. C. Lindsey and M. K. Simon, "Carrier Synchronization and Detection of Polyphase Signals," *IEEE Trans., COM-20*, pp. 441–454, June 1972.

15. M. K. Simon, "On the Calculation of Squaring Loss in Costas Loops with Arbitrary Arm Filters," *IEEE Trans., COM-26*, pp. 179–183, January 1978.

16. M. K. Simon, "Tracking Performance of Costas Loops with Hard-Limited In-Phase Channel," *IEEE Trans., COM-26*, pp. 420–432, April 1978.

17. R. J. Sherman, "Quadriphase Shift Keyed Signal Detection with Noisy Reference Signal," *EASCON 1969 Record*, pp. 45–52.

18. G. L. Hedin, J. K. Holmes, W. C. Lindsey, and K. T. Woo, "Theory of False Lock in Costas Loops," *IEEE Trans., COM-26*, pp. 1–11, January 1978.

19. M. K. Simon, "The False Lock Performance of Costas Loops with Hard-Limited In-Phase Channel," *IEEE Trans., COM-26*, pp. 23–34, Jan. 1978.

20. A. J. Viterbi, *Principles of Coherent Communications*, McGraw-Hill, New York, 1966, p. 289.

21. C. R. Cahn, "Improving Frequency Acquisition of a Costas Loop," *IEEE Trans., COM-25*, pp. 1453–1459, December 1977.

22. W. R. Bennett, "Statistics of Regenerative Digital Transmission," *BSTJ*, Vol. 37, pp. 1501–1542, November 1958.

23. P. A. Wintz and E. J. Luecke, "Performance of Optimum and Suboptimum Synchronizers," *IEEE Trans., COM-17*, pp. 380–389, June 1969.

24. J. J. Stiffler, *Theory of Synchronous Communications*, Prentice-Hall, Englewood Cliffs, NJ, 1971, Chap. 7.

25. U. Mengali, "A Self Bit Synchronizer Matched to the Signal Shape," *IEEE Trans., AES-7*, pp. 686–693, July 1971.

26. W. C. Lindsey and M. K. Simon, *Telecommunication Systems Engineering*, Prentice-Hall, Englewood Cliffs, NJ, 1973, Chap. 9.

27. M. K. Simon, "Nonlinear Analysis of an Absoulte Value Type of an Early–Late Gate Bit Synchronizer," *IEEE Trans., COM-18*, pp. 589–597, October 1970.

28. W. C. Lindsey and R. C. Tausworthe," Digital Data-Transition Tracking Loops," *JPL SPS,37-50*, Vol III, pp. 272–276, Jet Propulsion Laboratory, Pasadena, CA, April 1968.

29. M. K. Simon, "Optimization of the Performance of a Digital-Data-Transistion Tracking Loop," *IEEE Trans., COM-18*, pp. 686–690, October 1970.

30. B. R. Saltzberg, "Timing Recovery for Synchronous Data Transmission," *BSTJ*, Vol. 46, pp. 593–622, March 1967.

31. F. M. Gardner, *Clock and Carrier Synchronization: Prefilter and Antihang-up Investigations*, ESA CR-984, European Space Agency, Noordwijk, Netherlands, November 1977.

32. J. E. Mazo, "Jitter Comparison of Tones Generated by Squaring and by Fourth-Power Circuits," *BSTJ*, Vol. 57, pp. 1489–1498, May–June 1978.

33. L. E. Franks and J. P. Bubrouski, "Statistical Properties of Timing Jitter in a PAM Timing Recovery Scheme," *IEEE Trans., COM-22*, pp. 913–920, July 1974.

34. Y. Takasaki, "Timing Extraction in Baseband Pulse Transmission," *IEEE Trans., COM-20*, pp. 877–884, October 1972.

35. E. Roza, "Analysis of Phase-Locked Timing Extraction Circuits for Pulse Code Transmission," *IEEE Trans., COM-22*, pp. 1236–1249, September 1974.

36. D. L. Duttweiler, "The Jitter Performance of Phase-Locked Loops Extracting Timing from Baseband Data Waveforms," *BSTJ*, Vol. 55, pp. 37–58, January 1976.

37. A Papoulis, *Probability, Random Variables, and Stochastic Processes*, McGraw-Hill, New York, 1961, Chap. 12.

38. C. R. Cahn, D. K. Leimer. C. L. Marsh, F. J. Huntowski, and G. D. Larue, "Software Implementation of a PN Spread Spectrum Receiver to Accomodate Dynamics," *IEEE Trans., COM-25*, pp. 832–840, August 1977.

39. U. Mengali, "Joint Phase and Timing Acquisition in Data-Transmission," *IEEE Trans., COM-25*, pp. 1174–1185, October 1977.

40. W. R. Bennett and J. R. Davey, *Data Transmission*, McGraw-Hill, New York, 1965, Chap. 5.

Appendix A

Mathematical Review

This appendix is a collection of some of the underlying mathematics used in analysis of phaselock loops. Since it is intended solely as a brief review and not as a rigorous development, the pertinent results are stated without proof. References are provided for the reader who wants greater detail.

A.1 NETWORK ANALYSIS

Laplace Transforms

It is convenient to make use of the concepts and notation of Laplace transforms. (The standard engineering reference on the subject is M. F. Gardner and J. L. Barnes, *Transients in Linear Systems*, Wiley, New York, 1942.) We denote the Laplace complex variable (sometimes called *complex frequency*) by the symbol $s = \sigma + j\omega$.

For any physical time function $x(t)$, such that $x(t) = 0$ if $t < 0$, there is a function $X(s)$, which is the Laplace transform of $x(t)$. Similarly, for every suitable complex function $X(s)$ there is a unique time function $x(t)$, which is the inverse Laplace transform of $X(s)$. These relations may be symbolized by the transform pairs

$$L[x(t)] = X(s)$$

$$L^{-1}[X(s)] = x(t)$$

Several transform pairs used in this book are tabulated at the top of the next page.

Transfer Functions

Let us consider a linear, constant-element, two-port, electrical network to which an input signal $x_{in}(t)$ is applied and which, in response, delivers

252

an output signal $x_o(t)$ as in Figure A.1. The Laplace transforms of the signals are $X_{in}(s)$ and $X_o(s)$ respectively.

	$x(t)$	$X(s)$
Unit step	1	$\dfrac{1}{s}$
Unit ramp	t	$\dfrac{1}{s^2}$
Exponential	e^{-at}	$\dfrac{1}{s+a}$
Derivative	$\dfrac{dy(t)}{dt}$	$sY(s)-y(0+)$
Definite integral	$\displaystyle\int_0^t y(t)\,dt$	$\dfrac{1}{s}Y(s)$
Final value	$\displaystyle\lim_{t\to\infty} x(t)$	$\displaystyle\lim_{s\to0} sX(s)$

The transfer function of the network is defined as

$$F(s) = \frac{X_o(s)}{X_{in}(s)}$$

For all purposes of this book, $F(s)$ is the ratio of two polynomials in s.

$$F(s) = \frac{a_m s^m + a_{m-1}s^{m-1} + \cdots a_o}{b_n s^n + b_{n-1}s^{n-1} + \cdots b_o}$$

($m \leqslant n$ for a realizable network.) A polynomial may be factored and expressed as a product of its roots.

$$F(s) = \frac{a_m(s-z_m)(s-z_{m-1})\cdots(s-z_1)}{b_n(s-p_n)(s-p_{n-1})\cdots(s-p_1)}$$

Roots of the numerator are called zeros, whereas roots of the denominator are called poles. There are m zeros and n poles. The network is said to be of nth order; the order is equal to the number of poles, which is the same as the degree of the denominator.

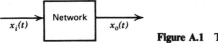

Figure A.1 Two-port network.

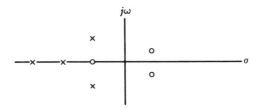

Figure A.2   Pole–zero diagram: (O) zeros; (×) poles.

Roots may be plotted in the complex s-plane as shown in Figure A.2. Complex roots must appear in conjugate pairs, whereas real roots may appear alone. Poles must appear in only the left half-plane (LHP) if the network is to be stable (realizable). Zeros may appear anywhere, but we are concerned only with networks whose zeros are in the LHP (minimum phase networks).

Frequency Response

The frequency response of a network—that is, the steady-state response to a sinusoidal input—is directly related to the transfer function as now described. Let us suppose the input to a network is of the form

$$x_{in}(t) = \sin \omega t$$

Then the steady-state output will be

$$x_0(t) = A(\omega) \sin \left[\omega t + \phi(\omega) \right]$$

where $A(\omega)$ is the amplitude of the output and $\phi(\omega)$ is the phase. A graph of A and ϕ plotted versus ω shows the frequency response of the network.

The transfer function of the network may be written for the condition $s = j\omega$ (i.e., $\sigma = 0$). Wherever s appears in the expression $F(s)$, we substitute $j\omega$ instead, thereby obtaining $F(j\omega)$. Written in polar form,

$$F(j\omega) = |F(j\omega)| \exp j \, Arg \, F(j\omega)$$

$$= A(\omega) e^{j\phi(\omega)}$$

That is, the magnitude of the transfer function for $s = j\omega$ is equal to the amplitude of the frequency response of the network, and the argument (phase) of the transfer function is equal to the phase of the frequency response.

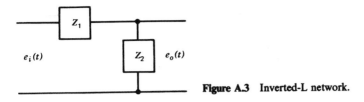

Figure A.3   Inverted-L network.

Examples

Several common networks are often encountered in phaselock circuits; their transfer functions and some properties of interest are presented here.

First we consider a generalized inverted-L network consisting of series and shunt impedance arms, as in Figure A.3. The transfer function is

$$F(s) = \frac{E_0(s)}{E_{in}(s)} = \frac{Z_2(s)}{Z_1(s) + Z_2(s)}$$

One particular L-network (Figure A.4) uses a resistor in the series arm and a capacitor as the shunt arm. Therefore, $Z_1 = R$, $Z_2 = 1/sC$, and the transfer function is

$$F(s) = \frac{1/sC}{R + 1/sC}$$

$$= \frac{1}{sCR + 1}$$

Frequency response, obtained by substituting $s = j\omega$, is

$$F(j\omega) = \frac{1}{j\omega CR + 1}$$

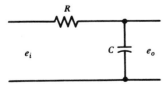

Figure A.4   RC lag network.

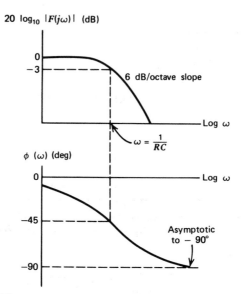

Figure A.5 Frequency response of lag network.

Magnitude of the frequency response is

$$|F(j\omega)| = [F(j\omega)F(-j\omega)]^{1/2}$$

$$= (1+\omega^2 R^2 C^2)^{-1/2}$$

and the phase is

$$\phi(\omega) = -\tan^{-1}\omega CR$$

It is often convenient to plot frequency response on a logarithmic frequency abscissa and also to plot magnitude on a logarithmic (decibel) ordinate. Phase is almost invariably plotted on a linear ordinate. Some properties of these graphs are shown in Figure A.5.

Another network, widely used as the loop filter in many phaselock loops, is the lag–lead network shown in Figure A.6. In this case $Z_1 = R_1$, $Z_2 = R_2 + 1/sC$ and

$$F(s) = \frac{R_2 + 1/sC}{R_1 + R_2 + 1/sC}$$

$$= \frac{sCR_2 + 1}{sC(R_1 + R_2) + 1}$$

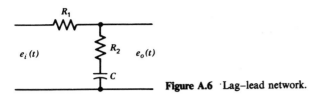

Figure A.6 ·Lag–lead network.

The frequency response is

$$F(j\omega) = \frac{1 + j\omega R_2 C}{1 + j\omega C(R_1 + R_2)}$$

$$|F(j\omega)| = \left\{ \frac{(1 + j\omega R_2 C)(1 - j\omega R_2 C)}{[1 + j\omega C(R_1 + R_2)][1 - j\omega C(R_1 + R_2)]} \right\}^{1/2}$$

$$= \left[\frac{1 + \omega^2 R_2^2 C^2}{1 + \omega^2 C^2 (R_1 + R_2)^2} \right]^{1/2}$$

$$\phi(j\omega) = \tan^{-1}\omega R_2 C - \tan^{-1}\omega C(R_1 + R_2)$$

and is sketched in Figure A.7.

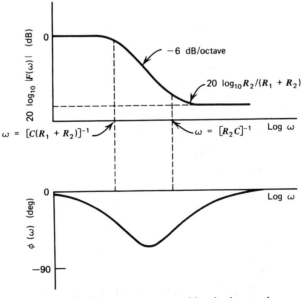

Figure A.7 Frequency response of lag–lead network.

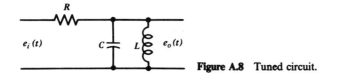

Figure A.8 Tuned circuit.

Another network of practical significance is the simple bandpass tuned circuit of Figure A.8. The transfer function is

$$F(s) = \frac{sL/R}{s^2LC + (sL/R) + 1}$$

from which the frequency response may be found to be

$$|F(j\omega)| = \frac{\omega L/R}{\left[(1 - \omega^2 LC)^2 + \omega^2 L^2/R^2\right]^{1/2}}$$

$$\phi(\omega) = \frac{\pi}{2} - \tan^{-1}\left(\frac{\omega L/R}{1 - \omega^2 LC}\right)$$

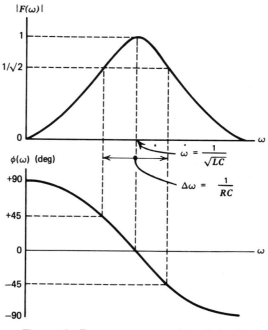

Figure A.9 Frequency response of tuned circuit.

Frequency response is plotted in Figure A.9 (note the linear frequency and magnitude scales).

The tuned circuit example is presented to introduce the concept of phase-slope. In any selective network the phase shift ϕ is necessarily variable as a function of the frequency ω. The variation of phase within the passband is referred to as the "phase-slope" of the response.

If information is carried in the phase of a signal and frequency of the signal is continually changing (e.g., the Doppler signal from a satellite), the phase-slope acts to distort the desired information. In a passive filter, selectivity and phase-slope are inextricably entwined; if the bandwidth is narrow, the phase-slope is necessarily steep. An important advantage of a phaselocked loop lies in the fact that narrow bandwidth can be attained with very small phase-slope.

A.2 FEEDBACK

A phaselock loop is a feedback network, and a brief discussion of some aspects of feedback theory is presented here. For greater detail refer to any of the numerous books on control systems and servomechanisms (e.g., John G. Truxal, *Control System Synthesis*, McGraw-Hill, New York, 1955).

Basic Equations

A simple feedback loop is shown in Figure A.10. It consists of a forward-gain network with transfer function $A(s)$, a feedback network with transfer function $B(s)$, and a subtractor network that forms the difference between the input signal $y_i(t)$ and the output of the feedback network $y_f(t)$.

Writing the applicable equations in the transform domain yields

$$Y_d(s) = Y_{in}(s) - Y_f(s)$$

$$Y_o(s) = A(s)\, Y_d(s)$$

$$Y_f(s) = B(s)\, Y_o(s)$$

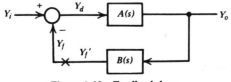

Figure A.10 Feedback loop.

The signals y_i, y_d, and y_o are known as the input, error, and output signals, respectively.

Combining the three equations for the individual components, we obtain the closed-loop transfer function

$$H(s) = \frac{Y_o(s)}{Y_i(s)} = \frac{A(s)}{1 + A(s)B(s)}$$

and the closed-loop error response

$$\frac{Y_d(s)}{Y_i(s)} = \frac{1}{1 + A(s)B(s)}$$

To determine open-loop gain, we break the loop at any point, say "x," and set $y_i = 0$. We apply a test signal y_f at the input side of the break and compute the resulting signal y_f' at the output side of the break. Open-loop gain is defined as

$$\frac{Y_f'(s)}{Y_f(s)} = -A(s)B(s)$$

In many cases (including phaselock loops) the transfer function of the feedback network $B = 1$. For this situation the closed-loop transfer function becomes

$$H(s) = \frac{A(s)}{1 + A(s)}$$

the error response is

$$\frac{Y_d(s)}{Y_i(s)} = \frac{1}{1 + A(s)} = 1 - H(s)$$

and the open-loop gain is simply $-A(s)$.

Stability

A feedback loop can oscillate if its open-loop gain exceeds unity and simultaneously its open-loop phase shift exceeds 180°. At least one of the closed-loop poles of an unstable loop will lie in the right half of the s-plane. Analysis of stability by investigating pole location is done in Chapter 2 by the root-locus method. To apply this technique, the loop

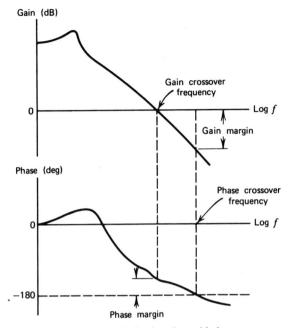

Figure A.11   Bode plot of a stable loop.

transfer function must be known in analytical form and the locations of the open-loop poles and zeros must be known.

Stability can also be analyzed, without knowledge of the pole locations or even an analytical description of the transfer function, by means of Bode plots. (H. W. Bode, *Network Analysis and Feedback Amplifier Design*, Van Nostrand, New York, 1945). A Bode plot consists of the pair of graphs of the magnitude* and phase of the frequency response of open-loop gain, $A(s)B(s)$. Magnitude and frequency are both plotted on logarithmic scales.

The Bode criterion for unconditional† stability is that the gain must fall below unity (0 dB) before the phase shift reaches 180°. Stable and unstable conditions are shown in Figures A.11 and A.12 respectively.

---

*Strictly speaking, in a Bode plot the magnitude is approximated by its asymptotes. We are free in our use of the terminology and let "Bode plot" include both the exact and approximate graphs of magnitude.

†A loop can be conditionally stable and still violate the Bode criterion.

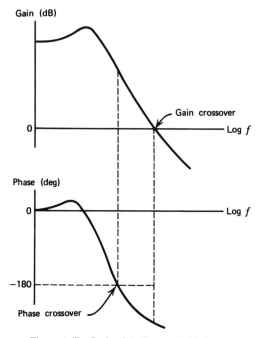

Figure A.12 Bode plot of an unstable loop.

A.3   NOISE FUNDAMENTALS

A major application of phaselock is to provide immunity against noise. This section briefly presents some of the more important concepts needed for the understanding of noise. Greater depth may be found in W. B. Davenport and W. L. Root, *Random Signals and Noise*, McGraw-Hill, New York, 1958 or A. Papoulis, *Probability, Random Variables, and Stochastic Processes*, McGraw-Hill, New York, 1965.

We now consider a real random noise voltage $n(t)$ and ask, "How may it be described for analytical purposes?" For random noise the actual waveshape is unpredictable, and it is not generally possible to write an explicit expression for $n(t)$. Any description must therefore be statistical in nature.

It is assumed that $n(t)$ is stationary; that is, all statistical properties are constant over all time. Most of the random quantities encountered in the study of phaselock may reasonably be considered stationary, although important nonstationary functions are encountered also.

Statistical Properties

Important statistical properties of noise are listed in the following paragraphs.

1. Probability density function (pdf) is denoted as $p(n)$. The integral $\int_{n_1}^{n_2} p(n)\,dn$ is the probability that the amplitude of a sample of $n(t)$ will lie in the range of n_1 to n_2. Any probability density $p(n) \geq 0$ for all n and

$$\int_{-\infty}^{\infty} p(n)\,dn = 1$$

2. Statistical averages.
   Mean value (average value, DC value):

$$E(n) = \int_{-\infty}^{\infty} n p(n)\,dn$$

where $E(n)$ is read as "statistical expectation of n."
   Mean square (intensity):

$$E(n^2) = \int_{-\infty}^{\infty} n^2 p(n)\,dn$$

Variance:

$$\sigma_n^2 = E(n^2) - E^2(n) = \int_{-\infty}^{\infty} \left[n - E(n)\right]^2 p(n)\,dn$$

3. Time averages.
   Mean value:

$$\overline{n(t)} = \lim_{T \to \infty} \frac{1}{2T} \int_{-T}^{T} n(t)\,dt$$

The bar over a quantity denotes time average. We are usually concerned with noise voltages having zero mean.
   Mean-square:

$$\overline{n^2(t)} = \lim_{T \to \infty} \frac{1}{2T} \int_{-T}^{T} n^2(t)\,dt$$

Variance:

$$\sigma_n^2 = \overline{n^2(t)} - \left[\,\overline{n(t)}\,\right]^2 = \overline{\left[n(t) - \bar{n}(t)\right]^2}$$

If the noise is *ergodic* (which includes stationary as a necessary condition) then the time averages and statistical averages are equal. In the body of the book it is assumed that all stationary functions are also ergodic and we do not distinguish between time and statistical averages. The overbar is used as a convenient symbol for either type of average, even where actual calculations would undoubtedly utilize the statistical expectation.

4. Gaussian probability density. A pdf found very often in nature and found even more frequently in analysis is the *gaussian* or *normal* density, given by

$$p(n) = \frac{1}{\sqrt{2\pi}\,\sigma_n} \exp\left[-\frac{(n-\bar{n})^2}{2\sigma_n^2} \right]$$

It has the familiar bell shape, as sketched in Figure A.13. Tails of the function extend to infinity in both directions. That is, there is some finite probability that a sample will exceed any arbitrarily large magnitude.

5. Autocorrelation. We define $n_1 = n(t_1)$, $n_2 = n(t_2)$, and $\tau = t_2 - t_1$. Also, $p(n_1, n_2)$ is the joint probability density of n_1 and n_2. The autocorrelation function is defined as

$$R(\tau) = E(n_1 n_2) = \int \int n_1 n_2 p(n_1, n_2)\, dn_1\, dn_2$$

If $n(t)$ is stationary, then $E(n_1 n_2)$ is also stationary; it is a function only

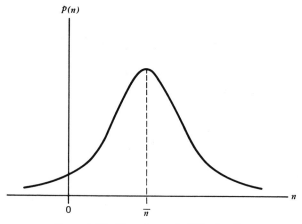

Figure A.13 Gaussian probability density.

of $\tau = t_2 - t_1$ and does not change with time. Autocorrelation is a very important quantity in its own right, but in this book it is used primarily as a vehicle to arrive at the next statistical property.

6. Spectral density. This is defined as the Fourier transform of the autocorrelation function.

$$\Psi_n(f) = \int_{-\infty}^{\infty} R(\tau) e^{-j\omega\tau} d\tau \qquad (\omega = 2\pi f)$$

The definition has meaning only for stationary autocorrelation. It is also true that

$$R(\tau) = \int_{-\infty}^{\infty} \Psi_n(f) e^{j\omega\tau} df$$

Spectral density is an extremely useful noise descriptor. It is an even function of frequency so that

$$\Psi_n(f) = \Psi_n(-f)$$

Spectral density integrated over all frequencies yields the intensity

$$\int_{-\infty}^{\infty} \Psi_n(f) df = \overline{n^2(t)} = R(0)$$

A complete description of $\Psi(f)$ may be obtained solely from its values at positive values of f. Thus, although mathematical definition of Ψ_n results in a *two-sided density*, it is also possible to speak of a *one-sided density* $\Phi_n(f)$, which involves only positive frequencies; that is,

$$\Phi_n(f) = 2\Psi_n(f) \qquad (f \geqslant 0)$$
$$= 0 \qquad (f < 0)$$

As defined here, dimensions of Φ_n are in V^2/Hz, so Φ_n is proportional to power. Spectral density could be defined in a slightly different manner, and the dimensions would then be in W/Hz. This second definition of power spectral density is denoted by the symbol $W(f)$ and in any particular case is related to $\Phi_n(f)$ by a constant multiplier.

Noise is often passed through filters. If input spectral density is $\Phi_i(f)$ and filter transfer function is $H(f)$, then the output spectrum is

$$\Phi_0(f) = \Phi_i(f) |H(f)|^2$$

A convenient fiction often employed is the concept of *white noise*. For this case $\Phi(f)$ is constant for all frequencies. No physical process can be truly white, for that would imply infinite power. A practical definition of whiteness is that the noise spectral density is constant at all frequencies of interest. A white noise spectrum is completely specified by a single number: the spectral density at any frequency. It is necessary to state one-sided or two-sided spectrum.

Caution. Noise is very commonly specified as white, gaussian noise. These are independent statements, and neither one implies the other. Noise can be nongaussian or nonwhite or both.

Noise Bandwidth

If we suppose a noise voltage has a one-sided spectral density of $\Phi(f)$ and that it is passed through a filter with transfer function of $H(j\omega)$, then intensity of the output voltage of the filter is

$$\overline{n_o^2} = \int_0^\infty \Phi(f)|H(j\omega)|^2 df$$

If the input spectrum is white, $\Phi(f) = N_o$ is constant for all f, and the output variance* becomes

$$\overline{n_o^2} = N_o \int_0^\infty |H(j\omega)|^2 df$$

At some reference frequency f_r, the transfer function has a magnitude of $H_r = |H(j\omega_r)|$.

There is an equivalent (though fictitious) rectangular filter, with constant magnitude of transmission equal to H_r and a bandwidth B_n (hertz), such that the variance of the outputs of the two filters are equal if they have white noise inputs of equal density; that is,

$$\overline{n_o^2} = N_o H_r^2 B_n$$

for the rectangular filter.

Equating the two variances provides the definition of noise bandwidth as

$$B_n = \frac{\int_0^\infty |H(j\omega)|^2 df}{|H(j\omega_r)|^2}$$

---

*If noise is white, there cannot be a DC component, so the mean is zero. Therefore, the intensity (mean square) is equal to the variance.

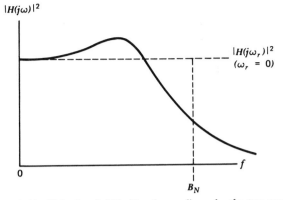

Figure A.14 Noise bandwidth. Equal areas lie under the two curves.

Figure A.14 shows the relation between an actual lowpass filter and its fictitious rectangular characteristic, which defines noise bandwidth.

It is usually convenient to take the reference frequency f_r to be zero in the case of a lowpass filter and as the center frequency in the case of a bandpass filter. Often a definition can be found wherein f_r is taken as the frequency of maximum response of the filter. Such a definition can lead to anomalous results when there is bandedge peaking of the response (as in Figure A.14) and therefore it is not used here.

A one-sided bandwidth has been defined, in accordance with normal convention and engineering convenience. It would have been equally feasible to define a two-sided noise bandwidth, which would, of course, have turned out to be $2B_n$ for the same filter. The one-sided definition is used throughout this book.

Noise Temperature and Noise Figure

Let us consider an amplifier or receiver that has a one-sided bandwidth B and is connected to a signal source of some kind (such as an antenna). Noise is generated within the amplifier (receiver) and also within the source. *Noise temperature* of the entire system is defined by attributing all the noise power generated to the source and considering the noise as thermal in origin, in which case the noise power referred to the source is given by

$$P_N = kT_eB$$

where k is Boltzmann's constant (1.38×10^{-23} joule/$^\circ$K), and T_e is the effective noise temperature of the system. The concept of noise temperature is most commonly applied in low-noise-receiving systems.

Noise figure is a measure of the noisiness of the amplifier as compared with that of the source. Specifically, noise figure is defined as the ratio of the total noise power delivered by the amplifier to that portion of the noise that is due to the source alone. For a thermal noise source the system noise power, referred to the source, is

$$P_N = kT_s BF$$

where T_s is the noise temperature of the source and $F \geqslant 1$ is the noise figure. (Usually $10 \log_{10} F$—the noise figure in decibels—is the number specified.)

It is common practice to define a standard noise figure with reference to a standard thermal source at temperature $T_s = T_o = 290°\mathrm{K}$ (i.e., slightly less than room temperature). For this condition $kT_o = 4 \times 10^{-21}$ W/Hz. If the source is not at T_o, the actual operating noise figure is not equal to the standard noise figure; however, the error in noise power computed with the standard noise figure and $T_s = T_o$ is small if the noise contribution of the receiver is much greater than that of the source. An amplifier noise temperature may also be defined as

$$T_A = (F-1)290°\mathrm{K}$$

where F is the standard noise figure.

Often it is useful to speak of the noise power spectral density W rather than the total power. Spectral density is given by $W = P_N/B$. In terms of system noise temperature $W = kT_e$, whereas in terms of standard noise figure $W = FkT_o$.

Narrowband Noise

If a gaussian noise voltage $n(t)$ has a relatively narrow bandpass spectrum, it is permissible and often opportune to write

$$n(t) = n_c(t)\cos\omega_1 t - n_s(t)\sin\omega_1 t$$

where ω_1 is any arbitrary frequency whatever, but simplest results are usually obtained if ω_1 is selected as being in the center of the narrow passband.

Some properties of this expansion are as follows:

1. Gaussian. If $n(t)$ is gaussian, n_c and n_s are also gaussian.

2. **Mean.** If $n(t)$ has zero mean, then n_c and n_s also have zero mean value. (A bandpass spectrum implies absence of DC and, therefore, zero mean for n.)

3. **Variance.**

$$\overline{n^2(t)} = \overline{n_c^2(t)} = \overline{n_s^2(t)}$$

4. **Independence.** The functions n_c and n_s are independent; that is,

$$\overline{n_c(t)n_s(t)} = \overline{n_c(t)}\,\overline{n_s(t)}$$

$$= 0 \quad \text{if} \quad \overline{n(t)} = 0$$

5. **Spectra.** We consider $n(t)$ to have a spectrum $\Phi_n(f)$. Then n_c and n_s have identical spectra of

$$\Phi_{ns}(f) = \Phi_{nc}(f) = \Phi_n(f_1 + f) + \Phi_n(f_1 - f)$$

(where f is always positive because we are using one-sided spectra). In essence, Φ_{nc} and Φ_{ns} are formed by translating Φ_n from f_1 to zero frequency, folding over the portion that translated past zero, and adding that to the unfolded portion, as shown in Figure A.15. The spectra Φ_{nc} and Φ_{ns} are lowpass functions if the arbitrary frequency f_1 is selected within the passband of Φ_n.

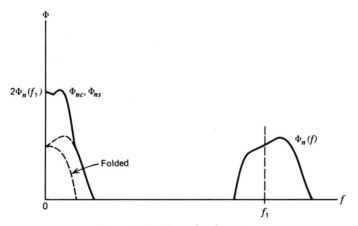

Figure A.15 Narrowband spectra.

6. **Complex Envelope.** The noise may be represented in polar form as

$$n(t) = A(t)\cos\left[\omega_1 t + \theta(t)\right]$$

where

$$A^2(t) = n_c^2(t) + n_s^2(t)$$

$$\theta(t) = \arctan\left[\frac{n_s(t)}{n_c(t)}\right]$$

The amplitude A is nonnegative and has a Rayleigh density

$$p(A) = \frac{A}{\sigma_n^2} \exp\left[-\frac{A^2}{2\sigma_n^2}\right]$$

with a mean value $\overline{A} = \sigma_n \sqrt{\pi/2}$ and mean square $\overline{A^2} = 2\sigma_n^2$. If the phase θ is considered to lie in the principle interval $(-\pi, \pi)$, then θ is uniformly distributed on the interval with density $1/2\pi$, mean value $\overline{\theta} = 0$ and variance $\overline{\theta^2} = \pi^2/3$. The density functions are shown in Figure A.16.

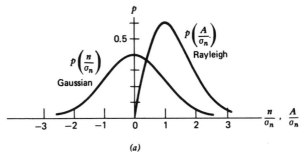

(a)

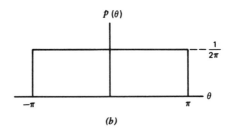

(b)

Figure A.16 Probability density functions: (a) amplitudes; (b) phase.

Appendix B

Analysis of Squaring Loop

Of all the different carrier synchronizer configurations, the squaring loop is the most tractable mathematically and its operation is the easiest to visualize. The oversimplified analysis presented here is intended to illustrate features that are common to all carrier synchronizers and to alert the reader to problems of analysis and operation. Our model is highly idealized; exact analysis of a realistic system has been an insuperable obstacle so far.

The squaring-loop configuration is shown earlier in Figure 11.2. We start with the voltage $v_{in}(t)$ that comes out of the bandpass filter and proceed much as in Chapter 3. We represent the input voltage as

$$v_{in}(t) = m(t) V_s \cos(\omega_i t + \theta_i) + n(t) \qquad (B.1)$$

where $n(t)$ is the bandpass gaussian noise as described in Chapter 3 and Appendix A and $m(t)$ is the data modulation on the signal *after filtering*. The quantities V_s and $n(t)$ have dimensions of volts while $m(t)$ is taken as dimensionless.

Because of the filtering, $m(t)$ cannot be a rectangular waveshape, even if that was the original modulating waveform. The discontinuities are rounded into continuous transitions and the individual pulses must overlap to some extent. Even if the original modulated signal had a constant envelope, the filtered signal has amplitude variations and has an envelope null at each phase reversal. An example is sketched in Figure B.1.

Moreover, the filtered modulation is not necessarily a real function. If the filter is not symmetrical about the carrier frequency, $m(t)$ is complex; filter asymmetry induces phase variations. Normally the filter is designed to be symmetrical when the loop is locked and tracking the signal, but prior to lock the filter of a long loop is detuned from the correct frequency and is therefore asymmetrical with respect to the signal.

Amplitude and phase variations caused by the filter are often neglected, but their effect is not necessarily trivial, (see Chapter 11).

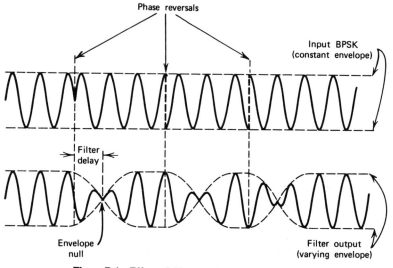

Figure B.1 Effect of filter on data-modulated signal.

Filtered input is applied to a squaring device whose operation is described by $v_{sq}(t) = K_{sq} v_{in}^2(t)$. (A square-law device is chosen because it is the easiest to handle mathematically, not because it is optimum in any sense. In fact, different nonlinearities can provide significantly improved performance.[1] An ordinary, full-wave, "linear" rectifier, for example, will yield about 2 dB less phase jitter of the recovered carrier at large input SNR.)

We expand the bandpass noise into two quadrature components about the carrier frequency, as in (3.3), and calculate v_{in}^2. As previously encountered with the phase detector, the squared output consists of low-frequency terms near DC and high-frequency terms around $2\omega_i$. Since we are trying to extract the carrier, we discard the low-frequency and consider only the double-frequency components. Pertinent output of the squarer is

$$v_{sq}(t) = K_{sq}\left[\tfrac{1}{2} V_s^2 m^2(t) \cos 2(\omega_i t + \theta_i) + \tfrac{1}{2} n_c^2 \cos 2\omega_i t \right.$$
$$- \tfrac{1}{2} n_s^2 \cos 2\omega_i t + n_c V_s m(t) \sin (2\omega_i t + \theta_i)$$
$$\left. + n_s V_s m(t) \cos (2\omega_i t + \theta_i) - n_c n_s \sin 2\omega_i t \right] \qquad \text{(B.2)}$$

Output from the VCO is represented as

$$v_o(t) = V_o \sin 2(\omega_i t + \theta_o) \qquad \text{(B.3)}$$

where, as in Chapter 3, we temporarily assume θ_o to be fixed, though random. Note that θ_o is the phase of the recovered carrier at ω_i but $2\theta_o$ is the phase of the VCO at $2\omega_i$.

Useful phase-detector output, discarding double-frequency products, is

$$v_d(t) = K_m v_{sq}(t) v_o(t) =$$

$$\tfrac{1}{2} K_m K_{sq} V_o \left[\tfrac{1}{2} V_s^2 m^2(t) \sin 2(\theta_i - \theta_o) + \tfrac{1}{2}(n_c^2 - n_s^2) \sin 2\theta_o \right.$$

$$- n_c n_s \cos 2\theta_o + n_c V_s m(t) \cos(\theta_i - 2\theta_o)$$

$$\left. - n_s V_s m(t) \sin(\theta_i - 2\theta_o) \right] \qquad (B.4)$$

We define:

$$\sigma_m^2 = \text{avg}\left[m^2(t) \right]$$

$$K_d = \tfrac{1}{4} K_m K_{sq} V_o V_s^2 \sigma_m^2$$

$$n_{c'} = \frac{n_c}{V_s \sigma_m}$$

$$n_{s'} = \frac{n_s}{V_s \sigma_m} \qquad (B.5)$$

We represent the squared modulation as $m^2(t) = m_o^2(t) + \sigma_m^2$, where $\text{avg}[m_o^2(t)] = 0$. Dropping the explicit dependence on t, the difference-frequency output of the phase detector becomes

$$v_d = K_d[\sin 2(\theta_i - \theta_o) \qquad \text{(DC error term)}$$

$$+ \left(\frac{m_o^2}{\sigma_m^2} \right) \sin 2(\theta_i - \theta_o) \qquad \text{(self-noise)}$$

$$+ (n_{c'}^2 - n_{s'}^2) \sin 2\theta_o - 2 n_{c'} n_{s'} \cos 2\theta_o \qquad \text{(noise} \times \text{noise)} \qquad (B.6)$$

$$+ 2(\frac{n_{c'} m}{\sigma_m}) \cos(\theta_i - 2\theta_o)$$

$$- 2(\frac{n_{s'} m}{\sigma_m}) \sin(\theta_i - 2\theta_o)] \qquad \text{(signal} \times \text{noise)}$$

Self-noise is usually neglected. If $\theta_i - \theta_o = 0$ or if $m_o^2 = 0$ (as would be true if the filtered pulses were rectangular), then the self-noise vanishes. In a real system, $m_o^2 \neq 0$, so there is an incentive for maintaining small tracking error.

The sum of the terms resulting from external noise is denoted $n'(t)$, so, discarding self-noise, the phase-detector output is

$$v_d(t) = K_d \left[\sin 2(\theta_i - \theta_o) + n'(t) \right]$$

$$\simeq K_d \left[2(\theta_i - \theta_o) + n'(t) \right] \tag{B.7}$$

where the second line represents the usual linearization.

Equations identical to (B.6) are obtained from an analysis of either the Costas loop or the remodulator of Chapter 11, provided that the lowpass filters are equivalent to the bandpass filter of the squaring loop and that the multipliers of the Costas or remodulator loops are ideal multipliers. If the loop dynamics are the same, the three loops yield identical performance. Any results obtained here are directly applicable to the other two configurations.

(Costas loops and remodulators often contain hard limiters, at least in part because a high-frequency multiplier that is linear on both input ports is difficult to construct. See Chapter 6 for discussion on switching phase detectors. Any circuit whose nonlinearity is other than pure square law is not exactly equivalent to the squaring loop. Results of squaring-loop analysis can be used as a good qualitative guide, but the actual amplitudes of phase fluctuations of the other circuits will not be identical to those of the squaring loop.)

We want to determine the VCO phase jitter caused by the equivalent noise $n'(t)$. To do this, we derive the one-sided spectral density $\Phi_{n'}(\omega)$ of n' and determine the phase variance as

$$\overline{(2\theta_{no})^2} = \frac{1}{2\pi} \int_0^\infty \Phi_{n'}(\omega) |H(j\omega)|^2 d\omega \tag{B.8}$$

much as is done in Chapter 3. To arrive at $\Phi_{n'}$, we first investigate the autocorrelation of $n'(t)$.

We start with the quantity $E[n'(t_1)n'(t_2)]$, where $t_2 = t_1 + \tau$ and $E[x]$ means "statistical average of x." We denote quantities at times t_1 and t_2 by subscripts 1 and 2 rather than writing out the functional time dependence. Then, assuming that message and noise are independent, remembering that n_c and n_s are also independent, and assuming θ_i and θ_o are fixed, the expectation is

$$E\left[n'_1 n'_2\right] = \sin^2 2\theta_o \left\{ E\left[n^2_{c'1} n^2_{c'2}\right] + E\left[n^2_{s'1} n^2_{s'2}\right] \right.$$

$$- E\left[n^2_{s'1}\right] E\left[n^2_{c'2}\right] - E\left[n^2_{c'1}\right] E\left[n^2_{s'2}\right] \right\}$$

$$+ 4\cos^2 2\theta_o \left\{ E\left[n_{c'1} n_{c'2}\right] E\left[n_{s'1} n_{s'2}\right] \right\}$$

$$+ 4\left(\frac{E\left[m_1 m_2\right]}{\sigma^2_m}\right) \left\{ \cos^2(\theta_i - 2\theta_o) E\left[n_{c'1} n_{c'2}\right] \right.$$

$$+ \sin^2(\theta_i - 2\theta_o) E\left[n_{s'1} n_{s'2}\right] \right\} \qquad \text{(B.9)}$$

The equivalent noises $n_{c'}$ and $n_{s'}$ are stationary, gaussian variables and have the same statistics. Accordingly,

$$E\left[n^2_{s'1}\right] = E\left[n^2_{c'1}\right] = E\left[n^2_{s'2}\right] = E\left[n^2_{c'2}\right] = \sigma^2_N$$

$$E\left[n_{c'1} n_{c'2}\right] = E\left[n_{s'1} n_{s'2}\right] = R_N(\tau) \qquad \text{(B.10)}$$

where σ^2_N denotes the variance of $n_{c'}$ or $n_{s'}$ and $R_N(\tau)$ denotes the autocorrelation. Furthermore, from the gaussian properties of the input noise, it can be shown that

$$E\left[n^2_{c'1} n^2_{c'2}\right] = E\left[n^2_{s'1} n^2_{s'2}\right] = \sigma^4_N + 2R^2_N(\tau) \qquad \text{(B.11)}$$

Since $\sin^2 x + \cos^2 x = 1$, substituting (B.10) and (B.11) into (B.9) gives

$$E\left[n'_1 n'_2\right] = 4R^2_N(\tau) + 4R_N(\tau)\frac{E\left[m_1 m_2\right]}{\sigma^2_m} \qquad \text{(B.12)}$$

which is independent of the phase angles, just as is found in Chapter 3.

It is tempting to identify (B.12) as the autocorrelation function of the noise, but $E[m_1 m_2]$ is not stationary and therefore does not have a corresponding spectral density. A further step is needed before transforming into the frequency domain.

Each data symbol is represented by a pulse. In most communications links, all pulses have identical waveforms—denoted $p(t)$—and differ only in the weighting applied to each pulse. For the filtered, binary modulation, the message can be described as

$$m(t) = \sum a_k p(t - kT) \qquad \text{(B.13)}$$

where $a_k = \pm 1$. We impose the restriction that the data be random, with zero mean, and separate bits are independent, that is,

$$E[a_k] = 0$$

$$E[a_k a_j] = 0 \quad \text{for} \quad j \neq k \tag{B.14}$$

Since $t_2 = t_1 + \tau$, substituting (B.13) for $m(t)$ gives

$$E[m_1 m_2] = E\left[\sum a_n p(t_1 - nT) \sum a_k p(t_1 + \tau - kT)\right]$$

$$= \sum \sum p(t_1 - nT) p(t_1 + \tau - kT) E[a_n a_k]$$

$$= \sum p(t_1 - kT) p(t_1 + \tau - kT) \tag{B.15}$$

which, by changing the dummy index, can readily be shown to be a periodic function of t_1, with period T. Bennett[2] has named this property *cyclostationary* and it is an important feature in the study of synchronous data streams.

To remove the time dependence, we obtain the time average of the statistical expectation. Because of the periodicity, the time average need only be computed over a T-sec interval. That is

$$\overline{E[m_1 m_2]} = \frac{1}{T} \int_0^T E[m_1 m_2] \, dt$$

$$= \frac{1}{T} \sum \int_0^T p(t - kT) p(t + \tau - kT) \, dt$$

$$= \frac{1}{T} \int_{-\infty}^{\infty} p(t) p(t + \tau) \, dt \tag{B.16}$$

where the overbar designates time average and the sum is for k from $-\infty$ to $+\infty$. The second line may be seen to be the infinite sum of adjoining integrals, which simplifies to the single infinite integral of the third line.

We recognize (B.16) as the conventional definition of autocorrelation of a random pulse train and give it the symbol $R_m(\tau)$. Using (B.16), we obtain the stationary autocorrelation of the noise output of the PD as

$$R_{n'}(\tau) = 4R_N^2(\tau) + \frac{4R_N(\tau) R_m(\tau)}{\sigma_m^2} \tag{B.17}$$

The two-sided spectral density of the noise is the Fourier transform of (B.17):

$$\Psi_{n'}(\omega) = \frac{4}{2\pi} \int_{-\infty}^{\infty} \Psi_N(\omega - \nu)\Psi_N(\nu)\,d\nu + \frac{4}{2\pi\sigma_m^2} \int_{-\infty}^{\infty} \Psi_N(\omega - \nu)\Psi_m(\nu)\,d\nu$$

$$(B.18)$$

and the one-sided density is $\Phi_{n'} = 2\Psi_{n'}$ for $\omega \geqslant 0$. The quantities Ψ_N and Ψ_m are the Fourier transforms of R_N and R_m, respectively.

To find the tracking jitter of the PLL, it only remains to substitute $\Phi_{n'}$, as calculated from (B.18), into (B.8). These formulas are broadly applicable and are subject primarily to the following restrictions:

- The PLL is assumed to operate in its linear region.
- The data stream is binary, with zero mean and with independent data values.
- The signaling pulses all have the same wave shape, which is a real function of time (implying a symmetrical bandpass input filter).
- Noise is gaussian.

Noise need not be white, the pulses need not be rectangular and, in fact, the pulses can even overlap, as they would in a narrowband channel; the analysis would still be valid.

There may well be applications in which one would be compelled to use the formulas as written, but physical insight requires simplification. First, let us assume that the bandwidth of the PLL is much smaller than that of the input filter; this condition is almost always met in practical systems. Then the jitter may be approximated by

$$\overline{(2\theta_{no})^2} = \Phi_{n'}(0)B_L \qquad (B.19)$$

since only the lowest-frequency components of the noise contribute significantly to the jitter.

As a further simplification, let us suppose that white noise is applied to the bandpass filter, which is assumed to have a rectangular passband with one-sided bandwidth B_i Hz centered on the carrier frequency. The one-sided noise density in the passband is taken as N_o (V^2/Hz). Accordingly, the two-sided spectral density of the equivalent baseband noise is (see

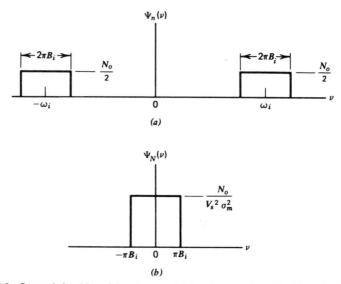

Figure B.2 Spectral densities of input noise: (a) bandpass noise $n(t)$; (b) equivalent base-band noise n_c or n_s.

Figure B.2)

$$\Psi_N(\nu) = \frac{N_o}{V_s^2 \sigma_m^2} \qquad |\nu| \leqslant \pi B_i$$
$$= 0 \qquad\qquad |\nu| > \pi B_i \qquad\qquad (B.20)$$

and, of course, Ψ_N is an even function.

Using that approximation, the first integral of (B.18) is evaluated to be $4N_o^2 B_i / V_s^4 \sigma_m^4$.

The next simplification is to assume that $m(t)$ has a rectangular wave shape—which is impossible at the output of a band-restrictive filter. Nonetheless, if B_i is much greater than $1/T$ (not the most effective choice of bandwidth), then the waveform might be approximated as rectangular. Under this assumption

$$R_m(\tau) = (1 - |\tau|/T) \qquad |\tau| \leqslant T$$
$$= 0 \qquad\qquad\qquad |\tau| > T \qquad\qquad (B.21)$$
$$\Psi_m(\nu) = T\left(\frac{\sin \nu T/2}{\nu T/2}\right)^2$$

The limits on the second integral of (B.18) would be $-\pi B_i$ to $+\pi B_i$, and the formal evaluation would be in terms of cumbersome sine integrals.

However, on the assumption that $B_i \gg 1/T$, the limits may be taken as $\pm\infty$ and the integral becomes $4N_o/\sigma_m^4 V_s^2$. Furthermore, the binary data values are ±1, so $\sigma_m^4 = 1$.

For these approximations, the zero-frequency spectral density of the noise evaluates as

$$\Psi_{n'}(0) = \frac{4N_o^2 B_i}{V_s^4} + \frac{4N_o}{V_s^2} \qquad (B.22)$$

and the phase jitter is

$$\boxed{\begin{aligned}\overline{(2\theta_{no})^2} &= 4\left(\frac{2B_L N_o}{V_s^2}\right)\left(1 + \frac{N_0 B_i}{V_s^2}\right) \\ &= 4\left(\frac{2B_L N_o}{V_s^2}\right)\left(1 + \frac{1}{2\rho_i}\right)\end{aligned}} \qquad (B.23)$$

where $\rho_i = V_s^2/2N_o B_i$ is the signal-to-noise ratio in the input filter bandwidth.

(Viterbi[3] and Stiffler[4] perform similar derivations, although the former is specialized to a half-sinusoidal pulse shape.)

REFERENCES

1. J. F. Oberst and D. L. Schilling, "The SNR of a Frequency Doubler," *IEEE Trans.*, COM-19, pp. 97–99, February 1971.

2. W. R. Bennett, "Statistics of Regenerative Digital Transmission," *BSTJ*, Vol. 37, pp. 1501–1542, November 1958.

3. A. J. Viterbi, *Principles of Coherent Communication*, McGraw-Hill, New York, 1966, Chap. 10.

4. J. J. Stiffler, "The Squaring Loop Technique for Binary PSK Synchronization," *SPS* 37-26, Vol. IV, pp. 240–246, Jet Propulsion Laboratory, Pasadena, CA, April 30, 1964.

Index